INTERNATIONAL SERIES OF MONOGRAPHS IN

THE SCIENCE OF THE SOLID STATE

GENERAL EDITOR: B. R. PAMPLIN

VOLUME 9

ELECTRICAL CONDUCTION IN SOLID MATERIALS

(Physicochemical Bases and Possible Applications)

ELECTRICAL CONDUCTION IN SOLID MATERIALS

(Physicochemical Bases and Possible Applications)

BY

J. P. SUCHET

Université René Descartes, Centre Luxembourg,

F-75270 PARIS CEDEX 06

PERGAMON PRESS

OXFORD · NEW YORK · TORONTO · SYDNEY · BRAUNSCHWEIG

PERGAMON PRESS LTD.,
Headington Hill Hall, Oxford

PERGAMON PRESS INC.,
Maxwell House, Fairview Park, Elmsford, New York 10523

PERGAMON OF CANADA LTD.,
207 Queen's Quay West, Toronto 1

PERGAMON PRESS (AUST.) PTY. LTD.,
19a Boundary Street, Rushcutters Bay, N.S.W. 2011, Australia

PERGAMON PRESS GmbH,
Burgplatz 1, Braunschweig 3300, West Germany

First edition 1975

Library of Congress Cataloging in Publication Data

Suchet, Jacques Paul, 1923–
Electrical conduction in solid materials.

(International series of monographs in the science of the solid state; v. 9)
Includes bibliographies and indexes.
1. Semiconductors. 2. Solids—Electric properties
I. Title.
QC611.S933 1974 537.6'22 74–19450
ISBN 0–08–018052–3

Printed in Great Britain by A. Wheaton & Co., Exeter

Suchet, so werdet ihr finden.
(Seek and ye shall find.)

(Lukas xi. 9)

By the Same Author

Les Varistances et leur emploi dans l'électronique moderne (*The Varistors and Their Use in Modern Electronics*) (in French), Chiron, Paris, 1955.

Chimie physique des semiconducteurs (*Physical Chemistry of Semiconductors*) (in French), Dunod, Paris, 1962; Russian translation, Izd. Metallurgiya, Moscow, 1964.

Id. (enlarged manuscript); English translation, van Nostrand, London, 1965; Polish translation, PWN, Warsaw, 1966.

Id. (again enlarged manuscript); Russian translation (2nd edn.), Izd. Metallurgiya, Moscow, 1969.

Curso de compostos quimicos semicondutores (*Course on Chemical Semiconducting Compounds*) (in French), Centro Técnico de Aeronautica, São José dos Campos, SP, Brazil, 1963.

Crystal Chemistry and Semiconduction in Transition Metal Binary Compounds, Academic Press, New York, 1971.

Edited by the Author

Séminaires de chimie de l'état solide (*Seminars on Solid State Chemistry*), yearly publication (in French):

1. *Liaisons interatomiques et propriétés physiques des composés minéraux* (*Interatomic Bonds and Physical Properties of Inorganic Compounds*), SEDES, Paris, 1969.
2. *Croissance de composés minéraux monocristallins* (*Growth of Monocrystalline Inorganic Compounds*), Masson, Paris, 1969; Russian translation, Izd. Metallurgiya, Moscow, 1970.
3. *Influence des changements de phase sur les propriétés physiques des composés minéraux* (*Influence of the Phase Changes on the Physical Properties of the Inorganic Compounds*), Masson, Paris, 1970.
4. *Appareillages et techniques de caractérisation des composés minéraux solides* (*Apparatus and Techniques of Characterization of the Solid Inorganic Compounds*), Masson, Paris, 1971.
5. *Quelques aspects de l'état solide organique* (*Some Aspects of the Organic Solid State*), Masson, Paris, 1972.
6. *Diagrammes de phases et stoechiométrie* (*Phase Diagrams and Stoichiometry*), Masson, Paris, 1973.
7. *L'Infrarouge en chimie des solides* (*The Infrared in Chemistry of Solids*), Masson, Paris, 1974.
8. *Les Solides divisés et dispersés* (*The Divided and Dispersed Solids*), Masson, Paris, 1974.
9. *Couches minces, émaux et vernis* (*Thin Layers, Enamels and Varnishes*), in preparation.
10. *Bilan et perspectives en chimie des solides* (*Review and Prospects in Chemistry of Solids*), in preparation.

Contents

Preface

THE local section of the Société Chimique de France, at the suggestion of Professor Aubry, invited me to deliver a lecture on 25 May 1972 at Nancy University entitled "Electrical conductivity in the solid state and its applications". I realized the scale of the subject on that occasion, and had regretted that the limited time at my disposal allowed me to give only a very rapid summary of the question.

Shortly afterwards, the Convention Intercantonale Romande pour l'Enseignement du 3ème Cycle en Chimie, at the suggestion of Professor Feschottes, invited me to give a course in Lausanne University, in February and March 1973, consisting of six lectures on "Semiconductor materials". This provided me with the opportunity I had wanted to deal in greater detail with conventional, magnetic, and switching semiconductor materials, and their various applications.

I then suggested to my friend Brian Pamplin that I should write up my notes, complete them by dealing also with conducting and insulating materials and their applications, and publish it all in the series "The Science of the Solid State", to provide students, teachers, and engineers —mainly those with chemical or metallurgical training—with a straightforward work linking the properties of such materials closely with their applications.

While British students often receive a balanced education as physico-chemists, this does not apply on the Continent, particularly in Germany, France, and Italy, where physics and chemistry are usually separated. Far too many physicists are thereby directed towards theoretical research, and many need retraining to be of any use in industry. This book will provide such students with a wider-ranging, simpler viewpoint than a treatise of strictly physical theory, the field of application of which is still so restricted.

As for an earlier work, the English text has been prepared in collaboration with Denis Mahaffey.

Paris J. P. SUCHET

Part I

Physicochemical Bases

Chapter 1

Conductors

1.1. Interatomic bonds

A solid phase contains a very large number of atoms, all identical in the case of an element, or of several different kinds in the case of an alloy or chemical compound. These atoms are linked with one another by means of their electrons on which cohesion of the solid depends. But these bonds can be of several different types, depending a great deal on the total ionization energies of the atoms present, in other words the energies needed to move all the valency electrons of these atoms to infinity. For an atom of a given element, the term of this operation is a positively charged ion, the electronic formula of which is that of the rare gas preceding the element in the Periodic Table of Elements. Table 1.1 gives these energies in electron volts for the main elements.

Atoms for elements which have low numbers (less than about 35) on this table combine with one another in the form of compact stacks of spheres, and part of their valency electrons escapes from the attraction of their nuclei, forming a gas of delocalized electrons ready to move in an electric field. This is the *metallic bond*, to be found in metals and alloys, where it gives rise to chemical formulae, the lower indices of which are usually neither simple nor unique for a given association of elements. Electrical conduction of phases with this type of bond is naturally high, and entirely due to the delocalized electron gas.

When one solid phase contains atoms of elements for which the numbers in Table 1.1 are in some cases lower and in other cases higher than 35, atoms in elements with numbers above 35 do not lose their electrons but tend to complete their electronic valency octet at the expense of atoms of elements with numbers below 35, thus attaining the

Table 1.1. BASED ON SUCHET, 1962

Li 5.4			Be 18.1		B 37.7	C 64.2	N 97.4	O 137	F 184
Na 5.1			Mg 15.0	Al 28.3	Si 44.9	P 64.7	S 87.6	Cl ?	
K 4.3	Cu 7.7	Ca 11.8	Zn 18.0	Ga 30.6	Ge 45.5	As 62.5	Se ?	Br ?	
Rb 4.2	Ag 7.5	Sr 11.0	Cd 16.8	In 27.9	Sn 39.4	Sb 55.5	Te (72)	I ?	
Cs 3.9	Au 9.2	Ba 9.9	Hg 18.6	Tl 29.7	Pb 42.1	Bi 55.7			

0 5 10 15 25 35 50 70 eV

Table 1.2. BASED ON SUCHET, 1962

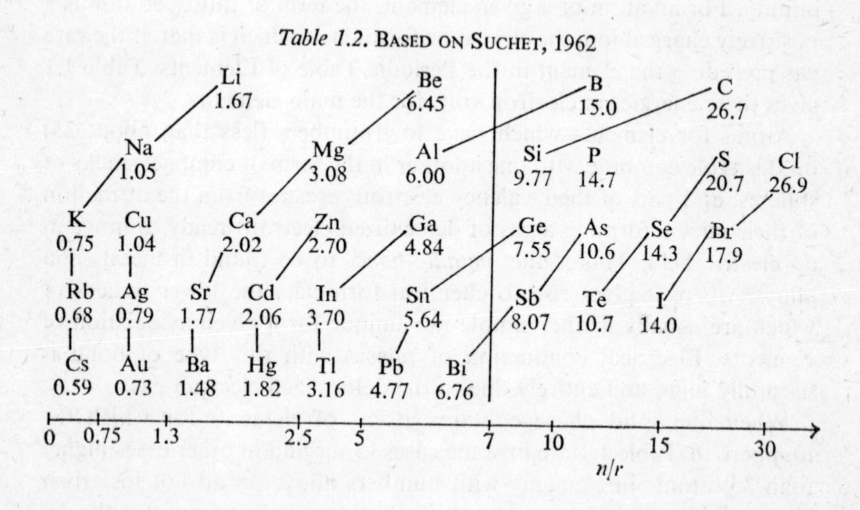

electronic formula of the rare gas following them in the Periodic Table of Elements. They therefore associate simply with one another, in accordance with known valency rules, in the form of stacks of charged spheres (cations and anions), at a distance such that their mutual attraction balances the repulsion of their electronic clouds. This is the *ionic bond*, to be found in many chemical compounds for which the formulae use low sub-indices, depending only on the columns of the classification table to which their elements belong. All the electrons are retained by the nuclei if the arrangement is perfectly regular, and electrical conduction is accordingly nil at very low temperatures.

This image of rigid spheres, however, soon proves false when the atomic number of the elements rises, together with the dimensions of the atom. Large ions can be deformed in the electric field created by their neighbours. The spherical symmetry of the electronic cloud of the atom is then no longer respected, and a dipole is induced inside the cloud. The electrical field responsible for this *polarization* of large ions is mainly due to highly charged small ions (cations). Table 1.2 shows the ratio of the number of elementary charges n carried by the cation to its ionic radius r for the main elements. This ratio is to some extent a measurement of the variation in relation to a pure ionic bond.

Let us assume that elements such as carbon, sulphur, or selenium are involved. Identical atoms cannot associate here by means of metallic bonds since the total ionization energies shown in Table 1.1 are too high. But they cannot associate either by means of an ionic bond since the n/r ratios shown in Table 1.2 are very high, and the very high polarization of the anions C^{4-}, S^{2-}, or Se^{2-} would immediately destroy such a bond. So what happens? The electronic clouds interpenetrate in the directions in which the cation field would be strongest, each atom pooling one electron in this direction with its neighbour. This is the *covalent bond*, the distinctive feature of which is accordingly its directional character.

In this pooling process, then, an atom cannot gain more electrons than it had to start with, and only atoms of elements in columns IV, V, VI, and VII of the Periodic Table can thus complete their octet by forming four shared electronic pairs per atom (tetrahedral structure of carbon atoms in the diamond), three pairs per atom (trigonal structure of phosphorus or arsenic atoms), two pairs per atom (right-angled spirated chain for selenium), or one pair per atom (Cl_2 molecules). All the electrons are either retained by the nuclei, as in an ionic bond, or immobilized in shared pairs, so that electrical conduction is still nil at very low temperatures.

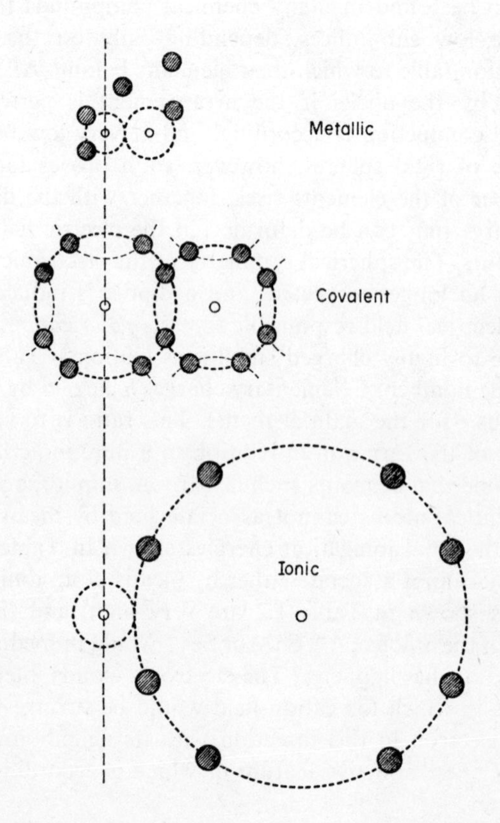

FIG. 1.1. Schema of the different types of interatomic bonds (based on Suchet, 1962).

The various types of bonds described above are shown in diagrammatical form in Fig. 1.1. For atoms of comparable mass, the interatomic distance increases as one passes from the metallic bond to the covalent bond, and then to the ionic bond.

1.2. Conductors and semiconductors

What happens to the electrical conduction if the field of very low temperatures is abandoned and the solid allowed to heat up? The effect of temperature on the electron gas mentioned in connection with the metallic bond is not very great, and *electronic conduction* accordingly remains largely unchanged. Let us leave aside the field of very low and low temperatures, to which we shall return in subsequent sections. A very simple, more or less linear, relation can be observed between conductivity σ and the centesimal temperature t:

$$\sigma \sim \sigma_0(1-0.004t),$$

where σ_0 is the conductivity at 0°C. The negative sign of the coefficient of t is explained by the thermal agitation of the atoms, which offers increasing opposition to the movements of free electrons. At room temperature, values range from around 5×10^5 mhos/cm for noble metals (Ag, Al, Au, Cu) to 10^4 mhos/cm for bismuth, with even lower figures for some transition metals such as manganese or lanthanides such as gadolinium.

In the case of a highly ionic bond, electrons remain attached to the nuclei until high temperatures are reached, and the solid is thus insulating. For instance, the conductivity of quartz SiO_2 is only 10^{-17} mhos/cm at room temperature. The diffusion coefficient of an element may, however, be sufficiently high in a compound, e.g. if the compound crystallizes in a defect structure for movements of ions to be possible. Very low conduction then appears, known as *ionic semiconduction*, which is always accompanied by electrolysis phenomena in the neighbourhood of the electrodes supplying the current, since material is being conveyed.* In the case of a covalent bond, elements with low atomic weights and their compounds remain insulating for a fairly long time, such as sulphur with only 10^{-16} mhos/cm and BN with 10^{-10} mhos/cm at room temperature. On the other hand, if there are elements with high atomic weight, breaking of some bonds is easier when temperature rises, releasing electrons that can conduct the electric current. This particular conduction, less than in metallic conduction, is known

* This phenomenon may be considered as an extension of the ionic conduction in *liquid* materials and, for this reason, will not be discussed in this book.

as *electronic semiconduction.* At room temperature, for instance, conductivity already reaches 10^{-4} mhos/cm for silicon, 10^{-1} mhos/cm for germanium, and 16 mhos/cm for InSb. There are even intermediate bonds in which ionic and electronic semiconductions coexist, e.g. in ZrO_2. In other words, while everything is fairly straightforward at very low temperatures, where solids either conduct or insulate, the situation becomes more complicated as the temperature rises.

The very slight ionic semiconduction often found in ionic, refractory insulating materials at high temperatures varies with temperature like atomic diffusion coefficients, in other words according to the relation $\exp(-Q/RT)$, where Q is expressed in cal/mole, $R = 1.99$ cal/degree (constant for perfect gases), and T is the absolute temperature. *Intrinsic* electronic semiconduction, found in relatively covalent insulators when they are pure, varies with temperature like the number of charge carriers released per cubic centimetre, in other words in accordance with the relation $\exp(-E/2kT)$, where E is expressed in electron volts, $k = 8.62 \times 10^{-5}$ eV/degree (Boltzmann's constant), and T is again the absolute temperature. Both phenomena, accordingly, are more or less related to temperature in the same way. This is because exponential laws occur frequently in physics, particularly as regards the effect of temperature. A very large number of solids follow this law, and are called semiconductors. Their common characteristic is that they are insulating at very low temperatures, and at room temperature still have a much lower conductivity than metallic conductors, not more than about 10–100 mhos/cm. It should be pointed out, however, that several widely differing semiconduction mechanisms can cause this behaviour.

If, instead of raising the temperature, the voltage applied between the two electrodes on the solid is increased, very different effects are once again obtained, depending on whether metallic conductors or insulating materials are involved. Let us assume that it is possible to keep the sample at the same average temperature, eliminating heat created by the Joule effect by means of an external cooling system. In the case of metallic conductors, nothing particular happens: current increases in proportion to voltage. In the case of insulating materials, on the other hand, much higher voltages have to be applied to obtain slight ionic or electronic semiconduction. A solid is never perfectly homogeneous, and the current thus produced will follow preferential paths, along which it will be much higher in intensity than if it were distributed uniformly throughout the volume. Since electrical insulators are generally also thermal insulators, a localized temperature rise caused by the Joule effect is to be expected, bringing about a drop in resistivity and unstable conditions, with the possibility of localized melting and breakdown. In any case, even if such localized overheating does not occur, e.g. when voltage increases very rapidly, a breakdown always happens when a level of about a megavolt per centimetre is reached. With one exception, which will be dealt with in Chapter 4, such a breakdown is always destructive in solids.

1.3. Metals and alloys

Let us now reconsider conducting solids—metals and alloys—in greater detail. Their electrical conducting properties are limited by the unavoidable causes of electron dispersion. There are many of these. At usual temperatures, the main cause is collisions with the lattice of atoms. In solids containing magnetic atoms, there is also the effect of their moments on the spin of the electrons. Finally, at low temperatures, impurities and structural defects play a predominant part. These three causes can be separated into three cumulative terms of resistivity. This is Matthiessen's rule, which can be expressed as:

$$\rho = \rho_{\text{atoms}} + \rho_{\text{spin}} + \rho_{\text{impurities}}.$$

Many theories have been put forward to explain the first term, and we shall try to summarize them here. To begin with, in 1900 Drude suggested that the electrons of a metal are completely free, and likened them to the atoms of a perfect gas complying with the known law $PV = RT$. He then expressed electrical resistance as proportional to mv/ne^2l, where m is the mass of the electron, v its average velocity, which depends on temperature, n the number of free electrons per unit of volume, and l the average free path. This theory satisfactorily reflects the constant relationship between thermal and electrical conductivities. Unfortunately, he deduces a law of variation based on $T^{\frac{1}{2}}$, which is not borne out by experiment. In 1928 Sommerfeld introduced the principles of the quantum theory, and replaced the Maxwell–Boltzmann statistics used in the model of perfect gases with Fermi–Dirac statistics. To put it simply, this means that the electron possesses an associated wave, the length of which is not indifferent, and that its energy can have only certain successive discrete values. Apart from this, the electrons are still free to move without being diffused in an ideal lattice presenting no imperfection. Resistance is expressed in the same way, but the meaning of v changes, and the variation in relation to temperature now rests on that of l, infinity at absolute zero and thereafter inversely proportional to T. Certain improvements were then made by Bloch, particularly connected with the electrostatic potential of the lattice.

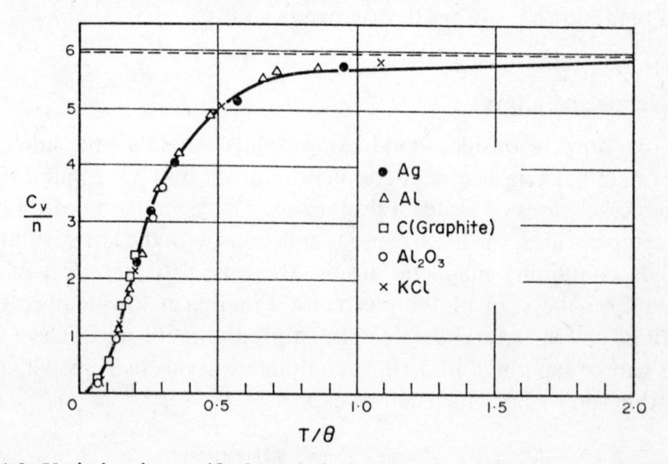

FIG. 1.2. Variation in specific heat (calories) in relation to the reduced absolute temperature T/θ for solids that obey Debye's theory (based on Seitz, 1943).

At temperatures above zero, the lattice of a crystal, even when pure and free from defects, is never ideal, since its atoms vibrate in relation to one another. These overall vibrations, known as *phonons*, propagate in the solid under conditions depending on the mass of the atoms, the ionicity of their bonds, and the temperature. They often constitute the main mechanism by which the solid can absorb heat, which explains the interest of the variation in their specific heat in relation to temperature, which tends towards a saturation value corresponding to Dulong and Petit's law. The temperature θ corresponding to a specific heat of less than 4% at saturation is taken as characteristic of the solid. Figure 1.2 shows the appearance of the variation in atomic heat in relation to the reduced temperature T/θ. This curve applies to solids as different as silver, graphite, and alumina, in which atomic vibrations constitute the main heat-absorption mechanism. Table 1.3 shows a number of other substances complying with this condition, with their temperature θ.

Table 1.3. BASED ON KITTEL, 1953, AND SEITZ, 1943

Metal	θ	Metal	θ	Metal	θ	Compound	θ
C diam.	1860	W	310	Ag, Hf	215	FeS_2	630
Be	1000	Re	300	Au	180	CaF_2	474
Cr	485	Mg, Ge	290	Sr	170	NaCl	281
Fe	420	Ir, Zr	280	Cd, Na	160	KCl	230
Ru, Al	400	Pd	275	Sb	140	AgCl	183
Co	385	Sn	260	Ga	125	KBr	177
Mo	380	Zn, Os	250	K, Hg, In, Tl, Bi	100	AgBr	144
Ni, Rh	370	Ta	245	Pb	88		
Mn, Ti	350	Ca	230	A	85		
Cu	315	Pt	225				

The characteristic temperature θ is usually called Debye's temperature, from the name of the person who defined it by means of an equation of the form $h\nu = k\theta$, where h is Planck's constant, ν the vibration frequency for the atoms, and k is Boltzmann's constant. Debye provided a particularly simple interpretation of the first term of Matthiessen's rule, showing that it varies like $T/M\theta^2$, in an element with atoms of mass M. Experiment usually bears out the law based on T. However, variations frequently occur at low temperatures. Figure 1.3 illustrates the case of silver.

When two metals can be mixed in a certain range of compositions, resistivity of the alloy varies evenly with composition, either lineally (Végard's law) or, more frequently, in accordance with a complex law. One of the oldest observations concerns Tl–Bi alloys, in which the γ phase contains 55–64% bismuth atoms. The Russian chemist Kurnakov used as an argument the absence of an extremum of electrical conductivity for the composition with 62.8% bismuth atoms (maximum melting temperature) to suggest a major exception to Dalton's laws, and forge the word "berthollide" (Fig. 1.4). In the case of interstitial compounds such as TiC_x, it is the density of free electrons that varies lineally in relation to composition, and is eliminated here for $x = 1$. It

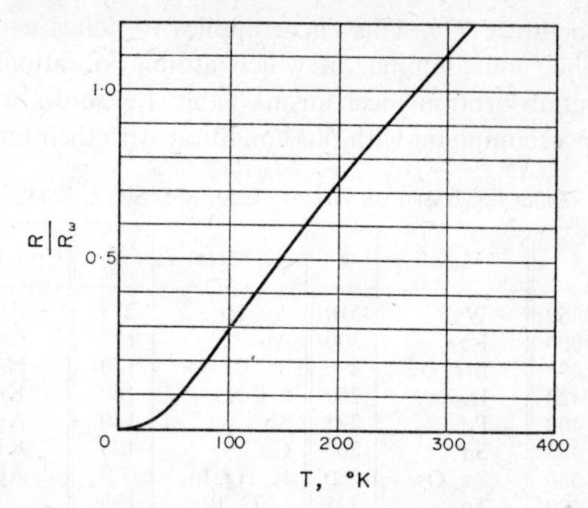

FIG. 1.3. Variation in resistivity of silver (ratio to the value at 0°C) in relation to absolute temperature (based on Seitz, 1943).

is difficult to give a general rule. One particular case consists of two metals that can give an ordered and a disordered alloy, with transition from order to disorder at a given temperature. Figure 1.5 shows the typical example of β brass.

1.4. Magnetic conductors

In certain elements—transition metals—incomplete filling of the layer of electrons gives the atoms a magnetic moment. In others—lanthanides—incomplete filling of the layer of f electrons produces the same result. These moments frequently operate in disordered directions (paramagnetism) and, when order intervenes, it may be antiparallel (antiferromagnetism) or parallel (ferromagnetism). Parallel orientation, the only one that causes a resulting moment within a macroscopic field, appears to be linked to long interatomic distances. For the series of transition metals, for instance, chromium, manganese, and γ iron, are within the boundary, while α iron, cobalt, and nickel are outside it. In Heusler's alloys, where the distances Mn–Mn are greater, ferromagnetism appears. For the other series of transition metals, all known elements are within the boundary peculiar to the series. For lanthanides, only Gd and Tb are ferromagnetic at normal temperatures. The Curie

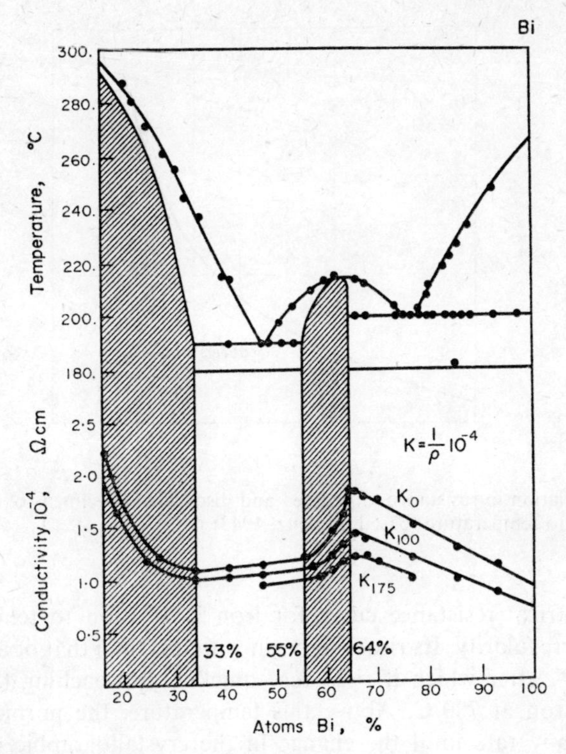

FIG. 1.4. Thallium–bismuth phase diagram with variation in electrical conductivity at three temperatures (based on Kurnakov, 1914).

point T_C is the temperature at which thermal agitation of atoms succeeds in destroying the magnetic order (see Table 1.4).

In conductors it has been seen that a number of electrons escape from the attraction of the nuclei and move freely in the solid. Because of this, redistribution generally occurs between the d shell, responsible for magnetism, and the outer shells p and s, so that the moment at saturation does not correspond exactly to what might be expected, in view of the degree of filling of the d shell of the atom. But once this redistribution has occurred, the moment varies in relation to temperature in exactly the same way for all substances in the same group. Figure 1.6 shows this by superimposing experimental points for iron, nickel, and cobalt.

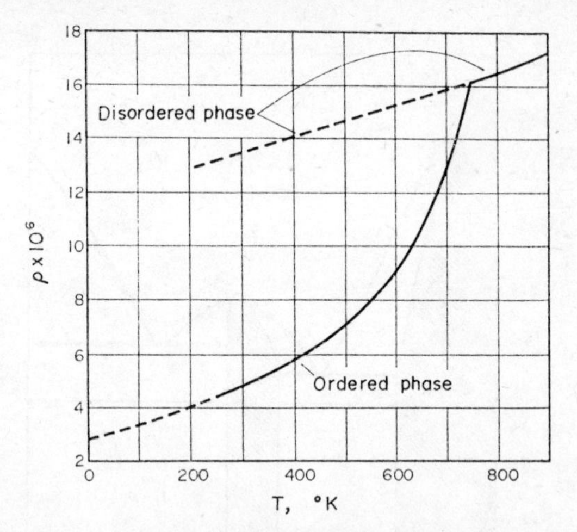

FIG. 1.5. Variation in resistance of ordered and disordered specimens of β brass in relation to temperature (based on Seitz, 1943).

The electrical resistance curve for iron in relation to temperature shows an irregularity. Its rise is to begin with less than that of a relation based on T, after which it rises very quickly, approaching the Curie point for iron at 780°C. Above this temperature, the normal law is found, at any rate until the change in the crystallographic structure α–γ. This property is shared by all ferromagnetic metals, and Becker and Döring, in a book published shortly before the last war, compared the reduced resistivities of nickel and its homologue in the second series, paramagnetic palladium, on the same graph. Figure 1.7 reproduces

Table 1.4. BASED ON MEADEN, 1965

Metal	T_C (°C)	Metal	T_C (°C)
Fe	780	Gd	20
Co	1075	Tb	−55
Ni	365	Dy	−188
		Ho, Er, Tu	−253

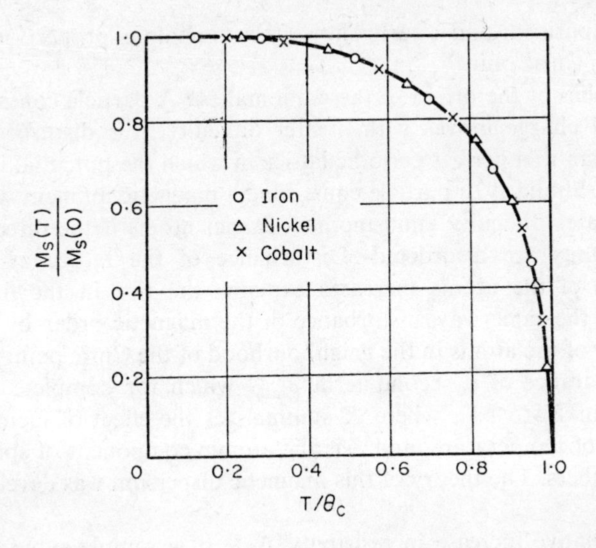

FIG. 1.6. Variation in ratio of the saturation magnetization at temperatures T and 0 in relation to the reduced absolute temperature T/T_C (based on Seitz, 1943).

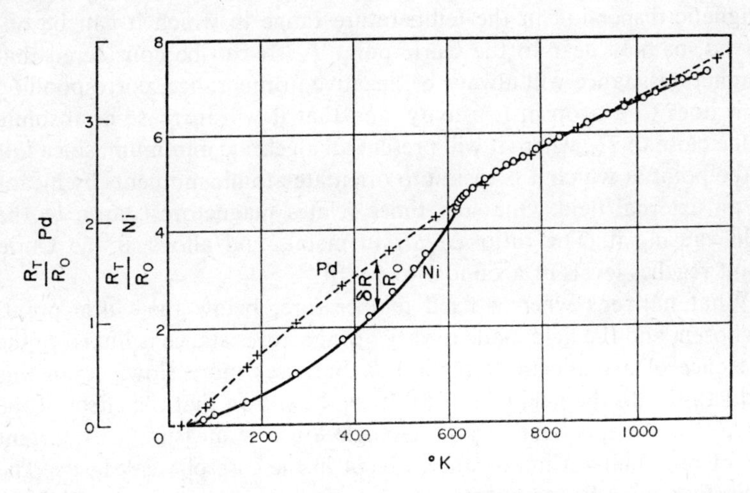

FIG. 1.7. Comparison of the variations in resistivity of nickel and palladium in relation to absolute temperature (based on Becker and Döring, 1939).

this comparison, and clearly shows the anomalous property of nickel below its Curie point.

What lies at the origin of these anomalies? A particle containing an electrical charge moves with greater difficulty in a disturbed crystal lattice than in a perfect periodic lattice in which the potential is strictly periodic. Similarly, a particle containing a magnetic moment will move with greater difficulty amid moment-carrier atoms if the directions of these atoms are disordered. Disturbance of the lattice as thermal agitation of the atoms increases explains the rise in the first term ρ_{atom}. In the same way, disturbance of the magnetic order by thermal agitation of the atoms in the neighbourhood of the Curie point explains the appearance of a second term ρ_{spin}, which for complete disorder will attain $KS(S+1)$, where K summarizes the effect of factors independent of temperature, and S is the atomic component of spin quantum numbers. The theory of this magnetic dispersion was developed by Kasuya.

The relative increase in resistivity $\Delta\rho/\rho$ of a sample subjected to a magnetic field is known as magnetoresistance. The effect of the field tends to reduce the disorder in the directions of the magnetic moments of the atoms by orientating them all parallel to itself. It thereby reduces magnetic dispersion in the temperature range in which it can be observed, namely near to the Curie point T_C. It can be considered that magnetoresistance will always be negative in this range, corresponding as it does to a drop in resistivity, and that it will increase in absolute value close to T_C, where it will present an algebraic minimum, since this is the point at which it is easiest to orientate atomic moments by means of an external field. One sometimes relates magnetoresistance to the field causing it. The ratio $\Delta\rho/\rho H$ in metals and alloys at the Curie point reaches levels of around 1 to 5×10^{-3}.

What happens when a fixed temperature, below the Curie point, is chosen, and the field made to vary? Magnetoresistance is linked to the existence of a magnetic induction B, but rises more slowly than this induction with the field (Fig. 1.8). It can be shown that the effect of the intensity of magnetization and its angle α to the measurement current are of predominant importance, except in the case of weak fields. The resistivity of a ferromagnetic domain can then be represented by a development in series related to $\cos\alpha$. Uneven powers are absent

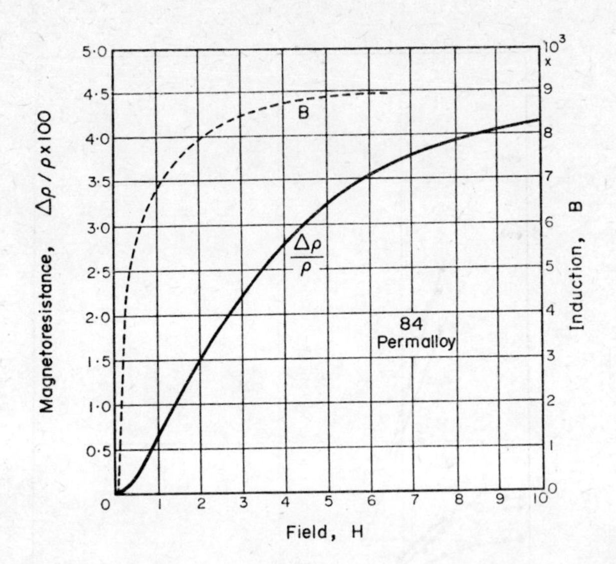

FIG. 1.8. Variation in magnetoresistance and magnetic induction of 84 Permalloy in relation to the applied field (based on McKeehan, 1930).

because of the symmetry of the effect, so that the magnetoresistance is found to be proportional to $(B-H)^2$ and, once the material is saturated, is found to decrease lineally.

Because of these effects, the resistivity of magnetic conductors can vary significantly from one material to another, and in a single material from one temperature to another. When a phase change causes an alteration in the magnetic order, the resistivity variation can occasionally be sudden and large (cf. Chapter 6).

1.5. Resistance at low temperatures

Of the three terms of Matthiessen's rule, as written above, it has been seen in turn that the first (ρ_{atoms}) disappears at absolute zero and the second (ρ_{spin}) is nil in the absence of magnetic atoms. The third term is sometimes referred to as the residual term because, in a non-magnetic metal, such as silver or aluminium, it is what remains when close to very low temperatures—in fact on reaching 4.2°K, the temperature of liquid

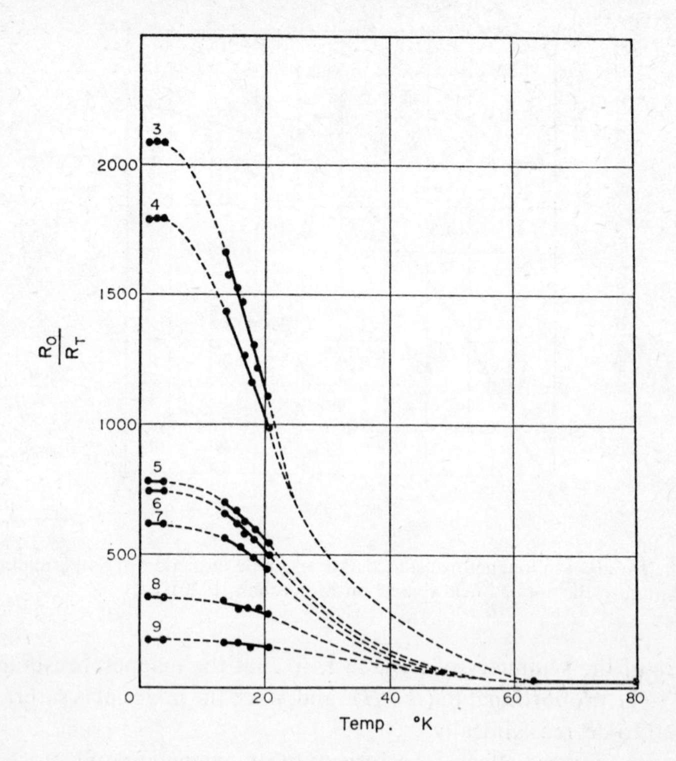

FIG. 1.9. Comparison of the variations in conductance (ratio to the value at 0°) of several aluminium samples, in relation to absolute temperature (based on Caron, 1957): (3) 99.9975%, (4) 99.9965%, (5–6) 99.992%, (7) 99.99%, (8) 99.98%, and (9) 99.965%.

helium. But it should also be pointed out that the proportional relationship of the first term to T is no longer valid at low temperatures, being replaced by a proportional relationship to T^a, where $3 < a < 5$. Figure 1.9 illustrates this law of variation, for samples of aluminium of different degrees of purity, by means of measurements of resistance close to the temperatures of liquid helium, liquid hydrogen, and liquid nitrogen. It will be seen that the resistance remains constant between 2.2 and 4.2°K, the extreme temperatures for measurements in liquid helium, and thus corresponds to Matthiessen's residual term.

For a non-magnetic metal, the expression "ideal resistance" is used to describe the difference between resistance measured at temperature T and the residual resistance measured at the temperature of liquid helium. The lower this residual term—in other words the purer the metal—the more accurate will be the measurement of this difference. Caron has studied aluminium samples of different degrees of purity and has found an exponent a of 3.1, regardless of the level of purity, between 14 and 20°K (3.7 for iron). Only the residual term is affected by impurities, the atoms of which restrict the average free path of the electrons, and it can be regarded as the sum of terms related to each of them, proportional to the number of atoms of impurity per cubic centimetre of host metal, and to a specific coefficient dependent on the difference between the atomic radius of the impurity and that of the host. Lattice defects, such as dislocations resulting from cold working, have a similar effect.

The effect of impurities at low temperatures has been mentioned first because it is of very general value, justifying the existence of a special term in Matthiessen's rule. But it is also because its investigation led historically to the discovery of superconduction. It was with the aim of obtaining a more easily purifiable metal that Kammerlingh Onnes, in 1911, thought of measuring the resistance of a mercury wire in helium, which he had succeeded in liquefying 3 years before. He soon noticed that the disappearance of all electrical resistance at very low temperatures was in fact in no way related to purity, occurred only in certain metals and alloys, and was not accompanied by any change in other properties. The change to the superconducting state occurs abruptly, at a characteristic temperature T_s, and can be eliminated by a characteristic value H_s of an outside magnetic field, with

$$H_s = H_0[1 - (T/T_s)^2].$$

This means that the current that can traverse a superconductor cannot exceed a level corresponding to the creation of a field H_s.

What microscopic and structural properties produce superconduction? For elements, Table 1.5 shows the position of elements which are superconducting in liquid helium in the Periodic Table. It will be seen that these elements are closely grouped, and appear to correspond simultaneously to two criteria: firstly, the possibility of multiple valencies 2, 3, and 4, or at least 2 and 4, and, secondly—with the exception of vanadium—a high atomic mass, corresponding to the second and third long periods. The second criterion has been confirmed by comparative study of the various isotopes of lead and tin, showing that T_s varies like the square root of the atomic mass. Superconducting compounds include NbB and $BaBi_3$ (6°), MoC (8°), ZrN (9.5°), NbC (10°), MoN (12°), and NbN (15°K). Finally, numerous alloys on the market make use in particular of niobium, tin, and lead.

Table 1.5. Based on Schoenberg, 1952

Li	Be											B	C	N	O	F
Na	Mg											Al	Si	P	S	Cl
K	Ca	Sc	Ti	V 5°1	Cr	Mn	Fe	Co	Ni	Cu	Zn	Ga	Ge	As	Se	Br
Rb	Sr	Y	Zr	Nb 8°	Mo	Tc	Ru	Rh	Pd	Ag	Cd	In 3°37	Sn 3°73	Sb	Te	I
Cs	Ba	La 4°37	Hf	Ta 4°4	W	Re	Os	Ir	Pt	Au	Hg 4°15	Tl 2°38	Pb 7°22	Bi	Po	At
Fr	Ra	Th	Pa	U												

Geller emphasizes the effect of certain privileged structures, such as that of βW (notably for Nb_3Sn) and the B1 structure of rock salt (notably for the compounds III VI, IV VI, and V VI), when interatomic distances are short, indicating a relatively covalent aspect in the bonds. The existence of two different valencies of the same atom, generating electronic exchanges, appears to be connected with superconductivity, e.g. in cubic $\square_x In_{1-x} Te$ (T_s would tend to reach zero for the semiconductor $\square In_2Te_3$, in which there is no more monovalent indium), SnAs (where there is as much divalent as tetravalent tin), $\square_x Ge_{1-x} Te$ and $\square_x Sn_{1-x} Te$. The density of carriers in a binary superconductor thus depends on the stoichiometric ratio of the components, and in many cases its calculation or measurement helps to explain variations in T_s depending on composition.

References

Becker, R. and Döring, W. (1939) *Ferromagnetismus*, Springer, Berlin.
Caron, M. (1957) *Publ. Scient. Techn. Ministère Air, Paris*, SDIT-328.

GELLER, S. (1969) in *Khimicheskaya sviaz' v poluprovodnikax i kristallax* (*The Chemical Bond in Semiconductors and Crystals*) (ed. N. N. Sirota), Nauka i Tekhnika, Minsk (English translation, Consultants Bureau, New York, 1971).
KITTEL, C. (1953) *Introduction to Solid State Physics*, Wiley, New York.
KURNAKOV, N. S. (1914) *Z. anorg. allgem. Chemie* **88,** 109.
MCKEEHAN, L. W. (1930) *Phys. Rev.* **36,** 948.
MEADEN, G. T. (1965) *Electrical Resistance of Metals*, Plenum, New York, 1965; Heywood, London, 1966.
SCHOENBERG, D. (1952) *Superconductivity*, Cambridge Univ. Press.
SEITZ, F. (1943) *Physics of Metals*, McGraw-Hill, New York and London.
SUCHET, J. P. (1962) *Chimie physique des semiconducteurs*, Dunod, Paris (English translation, van Nostrand, London, 1965).

Chapter 2

Conventional Semiconductors

2.1. Intrinsic semiconductors

Let us now return to the electronic semiconduction mentioned in § 1.2, confining ourselves for the moment to "conventional" semiconductors, namely inorganic elements and compounds not containing magnetic moment carrier atoms. Let us assume that all the electrons in the different atoms of the solid are classified on the basis of their energy. On the simplified graph in Fig. 2.1 they would occupy two possible regions, separated by a forbidden region of height E. Below would be the valency electrons providing chemical bonds between atoms, namely contained in the electronic octet of anions or shared Lewis pairs. Above would be electrons that have acquired sufficient energy to escape from the attraction of the nuclei and move freely in the crystal.

The electrical conduction activation energy E depends on the energy needed to break a chemical bond, and may be linked to the standard heat of formation of the crystal from its atoms. It may vary from about 0 to 10 eV. As the thermal activation energy kT increases, a statistically larger fraction of the valency electrons is released and can move under the influence of an electric field. Simultaneously, the empty places they leave behind them can also move. These are called "holes", and the same thing happens as if they were particles bearing an equal and opposite charge to that of the electron. This applies to a pure crystal without defects, in which charge carriers consist in equal numbers of electrons and holes. This is an "intrinsic" semiconductor.

Measurement of conductivity or resistivity in relation to temperature allows the value of E to be obtained quite easily. Points corresponding

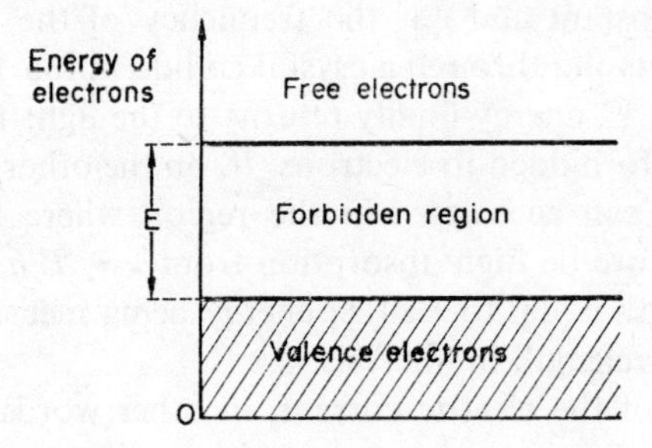

FIG. 2.1. Simplified energy diagram of the electrons in a crystal.

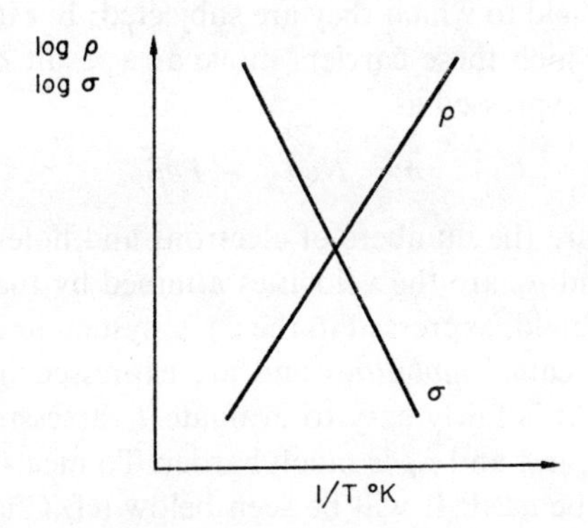

FIG. 2.2. Theoretical variations in logarithms of conductivity or resistivity in relation to the reciprocal of the absolute temperature.

to the different measurements are simply entered on a semilogarithmic graph as the reciprocal of the absolute temperature (Fig. 2.2). The equation

$$\sigma = \sigma_0 \exp(-E/2kT),$$

where σ_0 is the limit of conductivity at very high temperatures, becomes

$$\log \sigma = -(E/2kT)\log \sigma_0 \log e,$$

and the curve becomes a straight line with a slope $(E/2k)\log \sigma_0 \log e$. E can also be measured optically from the absorption edge of the light, which can be broken down into grains or photons of energy $h\nu$, where

h is Planck's constant and ν is the frequency of the electromagnetic wave. Photons passing through a crystal collide with a large number of electrons. If $h\nu < E$, energy finally returns to the light form, since this energy region is forbidden to electrons. If, on the other hand, $h\nu > E$, a valency electron can be excited in the region where it moves freely. There will therefore be high absorption from $\nu = E/h$, in other words from a wavelength equal to $1.24/E$, energy being measured in electron volts and the wavelength in microns.

The intensity of the electric current, in other words the number of elementary charges conveyed per second in a certain volume of solid, obviously depends on the number of carriers existing in this volume and the electric field to which they are subjected; but it also depends on the velocity at which these carriers move as a result of the field. Conductivity can be expressed as

$$\sigma = Ne\mu_N + Pe\mu_P,$$

where N and P are the numbers of electrons and holes per cubic centimetre and μ_N and μ_P are the velocities attained by these electrons and holes per unit of field, expressed in the c.g.s. system in cm/sec per V/cm. These ratios are called *mobilities* and are expressed in $cm^2/Vsec$. It is clear that, while it is fairly easy to evaluate E, assessment of the other parameters N, P, μ_N, and μ_P is much harder. To measure them, impure crystals have to be used. It will be seen below (cf. Chapter 6) that one of the types of carriers usually predominates in such crystals, so that half the variables can be eliminated by ignoring the other type. But this is not sufficient, and another equation is still needed. The simplest solution is generally to study the *Hall effect*.

This appears when magnetic induction in a perpendicular direction occurs in a substance traversed by an electric current, in other words usually when it is subjected to a magnetic field perpendicular to the direction of the current. Figure 2.3 shows that it reveals itself by a curving of the trajectory of the carriers, thereby accumulating static charges on surfaces A and B, parallel to the plane defined by the direction of the current and field. For convenience, two types of carriers have been separated: (*a*) shows positive charges, namely holes, and (*b*) shows electrons. The signs of the charges occurring on surfaces A and B are naturally opposed in these two cases.

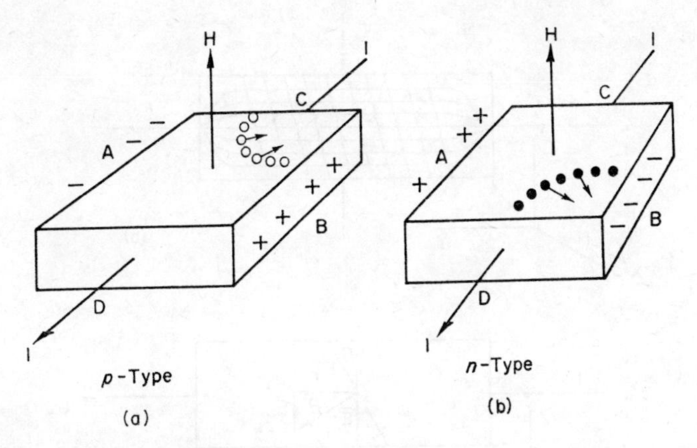

FIG. 2.3. Hall effect caused by magnetic field **H** on a current of holes (*a*) or electrons (*b*) in a parallelipiped *ABCD*.

The accumulation of charges on the side surfaces results in a modification of equipotential surfaces. Let us consider a cross-sectional example, perpendicular to the direction of the field, and let us assume that only one type of carriers is involved. Figure 2.4 shows, at (*a*), that the equipotential lines slope at an angle θ and that a potential difference, known as the Hall difference, appears between two electrodes that are symmetrical to the direction of the current. It is clear, at (*b*,) that this results from a rotation by the angle θ of the electric field vector **E**, which can then be broken down into E_y along the direction of the current, and E_{H}, the Hall component, along a perpendicular direction. Writing of the interactions between magnetic field and electric current shows that, in the case of a diamagnetic material, the angle θ is proportional to the magnetic field and electrical conductivity σ:

$$\theta = \mu H = R\sigma H.$$

The mobility of the carriers is accordingly seen as the ratio θ/H, being constant for a given sample and temperature, and it can also be expressed in the form of the product $R\sigma$, where R is a coefficient, again constant for a given sample and temperature, and inversely proportional to the number of carriers:

$$\mu = R\sigma = (1/Ne)\,(Ne\mu).$$

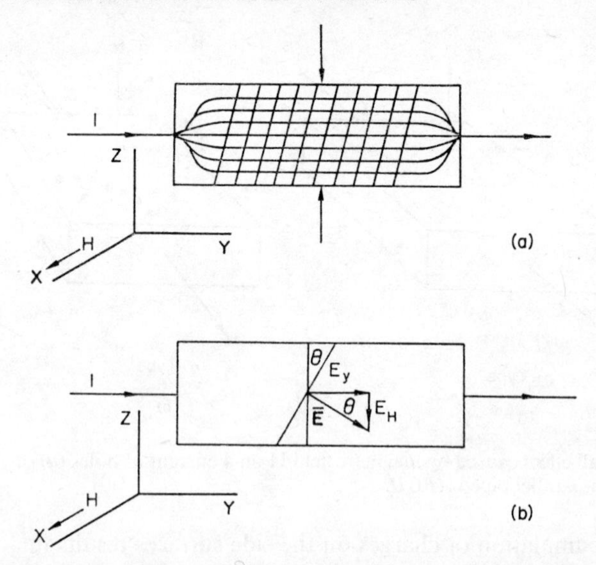

FIG. 2.4. (*a*) Rotation of an angle θ of equipotential lines by Hall effect. (*b*) Decomposition of the total electric field **E** into two perpendicular terms (based on Lindberg).

A sample with a coefficient R of 60 cm^3/C and conductivity σ of 1.5 mhos/cm will contain approximately 10^{17} carriers/cm^3, with a mobility in the region of 90 cm^2/V sec. Measurement of the Hall coefficient is therefore the best method of completing measurement of the resistivity of diamagnetic materials.

$E = 0.67$ eV at room temperature for germanium and 1.07 eV for silicon. $\mu_N = 3800$ cm^2/V sec for germanium and 1300 cm^2/V sec for silicon.

2.2. First condition

The electrical behaviour known as semiconduction does not appear at random in any chemical compound. It has been seen that it requires interatomic bonds to be at least partially covalent, and we might now go into slightly greater detail about this requirement. It in fact comprises two conditions. The first states that the number of available electrons on the atoms of the crystal at the highest level is just sufficient to form electronic bonding pairs, taking account of the orbital functions used in the spatial arrangements adopted by these atoms. The second condition states that the orbital functions are the *bonding* functions corresponding to (at least partly) covalent elements and compounds, and not the *antibonding* functions corresponding to metals and alloys.

The first condition is therefore essentially a crystallochemical condition, involving the crystallographic structure and numbers of electrons per atom. It was initially formulated 18 years ago by Mooser and Pearson, with special reference to the structure of blende:

$$n_e/n_a + b = 8,$$

where n_e was the total number of valency electrons corresponding to the stoichiometric formula, n_a the number of atoms of elements in columns IV to VII of the Periodic Table (atoms that can complete their electronic octet), and b the number of bonds that such atoms form directly with one another.

This rule is in fact derived from the rule attributed to Bradley and Hume-Rothery, according to which each atom in an element has a number b of immediate neighbours equal to $8 - a$, where a is the number of the column of the Periodic Table to which the element belongs, namely the number of its valency electrons. Each atom tends to complete its octet of s and p electrons by sharing with its neighbours in electronic bonding pairs. This means that

$$a + b = 8.$$

If one considers a binary semiconductor compound simply as a metalloid "inflated" by the contribution of the electrons of the metal, $a = n_e/n_a$, and one is back with Mooser and Pearson's rule. But only the valency electrons actually intervening in bonds should be counted. In the structure of blende, all the s and p electrons are involved, namely eight in all, whereas in rock salt only the p electrons—six in all—participate in bonding pairs. The first condition for the existence of semiconduction can therefore be expressed in its most general form as follows:

$$a + b = 2c,$$

where c is the number of electrons contributed by each atom to all the bonding pairs linking it to its closest neighbours (4 for sp^3 bonds in B3 blende, 3 for p^3 bonds in B1 rock salt, 2 for p^2 bonds in A8-type chains, etc.). In addition, the number of electrons counted at a must involve only atoms occupying metalloid sites in the crystallographic structure. These are easily determined by comparing the formula for the ternary or quaternary with that of the binary with the same structure.

Table 2.1 gives these rather dry explanations more immediacy by offering practical illustrations. For instance, for SiC, $n_e = 8$ and $n_a = 2$, so that $a = 4$ and there are 4 Si–C bonds between metalloids ($b = 4$). For ZnS, in contrast, $n_e = 8$ and $n_a = 1$ (S) so that $a = 8$ and there are no S–S bonds ($b = 0$). Finally, for $CdSnAs_2$, mere comparison of the formula for the ternary of chalcopyrite structure with that of the binary ZnS, with which the structure of blende is in fact identical, except for the alteration of cadmium and tin atoms on the zinc sites, shows that tin does not occupy metalloid sites. Accordingly $n_e = 16$, $n_a = 2$, $a = 8$, and, since there are no As–As bonds, $b = 0$.

For the binary compound PbS, with rock-salt structure, each atom has six neighbours, but forms only three electronic bonding pairs with them (there is said to be "resonance" of these three bonds among the six neighbours): $n_e = 6$, $n_a = 1$ (S), $a = 6$, and $b = 0$. For $TlSbS_2$, the wolfsbergite structure is not equivalent to that of PbS, even taking account of the alternation of thallium and antimony on the lead sites,

Table 2.1. BASED ON SUCHET, 1962a

$2c = 8$ (four bonding pairs or tetrahedral arrangement) s, p_x, p_y, p_z	Ge SiC ZnS $(CdSn)As_2$	$a = 4$ $a = 4$ $a = 8$ $a = 8$	$b = 4$ $b = 4$ $b = 0$ $b = 0$
$2c = 6$ (three bonding pairs or trigonal arrangement) p_x, p_y, p_z	As PbS $(TlSb)S_2$	$a = 3$ $a = 6$ $a = 6$	$b = 3$ $b = 0$ $b = 0$
$2c = 4$ (two bonding pairs or linear chain arrangement) p_x, p_y	S	$a = 2$	$b = 2$

but distortion is not considerable enough to remove the similarity in the two structures. It can be said that antimony does not occupy metalloid sites. $n_e = 12$, $n_a = 2$, $a = 6$, and $b = 0$.

2.3. Binary and ternary compounds

One particularly simple method of ensuring strict observance of the first condition of semiconduction is to write the chemical formula of a binary compound, deriving it from that of a semiconductor element by alternate isoelectronic substitution. In other words, the average number of valency electrons will remain constant. For an equiatomic binary AB, for instance, n_e will have to be equal to 8 in the tetrahedral arrangement and 6 in the trigonal arrangement. Let us leave aside the element boron, however, since it is complex in structure, and no known binary derivatives exist.

Carbon, in its diamond form, thus gives rise to III V, II VI, or I VII compounds, where I, II, III, V, VI, and VII are elements in columns with the same number in the Periodic Table, with 1, 2, 3, 5, 6, and 7 valency electrons. The average valency thus remains 4. The best known of these derivative compounds are the III V ones, discovered in Germany in 1955 by Welker. Table 2.2 gives a list of them, together with the corresponding semiconduction parameters E and μ_N. The order is that of increasing average atomic weights. It can be seen that, with a few exceptions (marked with an asterisk on the table), this order corresponds to the orders of decreasing E and increasing μ. We shall return to this point in the following section. II VI compounds mainly comprise those of the elements Zn, Cd, Hg and S, Se, Te. The only I VII compounds crystallizing in the structure of blende are CuCl, CuBr, CuI, and AgI. The effect of $(Z+Z')/2$ on semiconduction parameters is also felt.

Arsenic, in its A7 form, consists of double atomic layers with a trigonal arrangement not very different from the arrangement already mentioned for the structure B1 (cf. Fig. 2.5). It is therefore not surprising that it should produce, by the same method, compounds with a similar structure (SnS, structure B29) and compounds with the structure B1 (II VI or IV VI involving elements Ca, Sr, Ba, Sn, Pb and S, Se, Te). The average number of p electrons that can provide bonds therefore

Table 2.2. E AND μ BASED ON NEUBERGER, 1971

Formula	$(Z+Z')/2$	$E_{300°K}$	μ_N
AlN	10	5.9	?
AlP	14	2.45*	80
GaN	19	3.39	150
AlAs	23	2.9	180
GaP	23	2.78	2,100
InN	28	2.4	?
AlSb	32	2.218	200*
GaAs	32	1.428	16,000
InP	32	1.3511	44,000
GaSb	41	0.70	10,000*
InAs	41	0.356	120,000
InSb	50	0.18	1,000,000

* Exceptions to the order decreasing E and increasing μ_N.

remains 3 (in the case of column II elements, the two s electrons are excited at the p level). Sulphur, in its A8 structure form, does not appear to give binary derivatives.

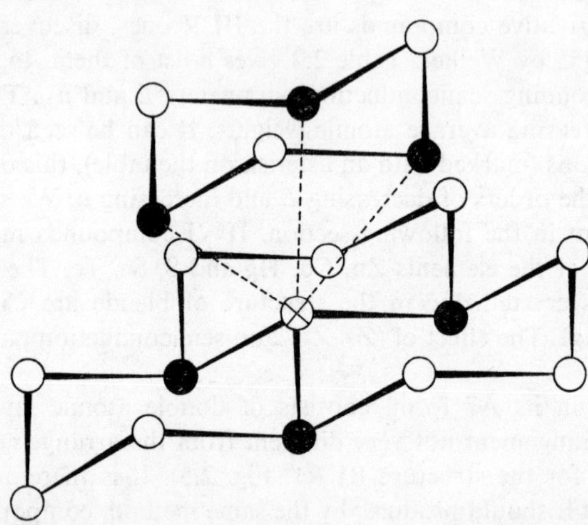

FIG. 2.5. Space configuration of the trigonal bonds of arsenic (around the atom marked with the cross) and, in dotted lines, their possible resonance (based on Krebs, 1968).

The resonance of the bonds of an atom between twice the number of neighbours, just encountered in the trigonal arrangement for binary derivatives with structure B1, can also occur, for only one of the components, in the tetrahedral arrangement. The A atom forming 4 sp^3 bonds is then situated in the centre of a cube, at the points of which are 8 B atoms. These B atoms can be grouped 4 by 4 in two tetrahedrons overlapping with each other. The structure thus defined corresponds to the structure of C1 fluorite (or antifluorite). It can easily be seen that isoelectronic substitution requires a formula AB_2 (or A_2B) such as II_2IV or I_2VI, in which elements Mg, Si, Ge, Sn, Li, Na, K, Ag and Se, Te can be involved.

Table 2.3. BASED ON SUCHET, 1962a

Formula								Structure	Example
		II		IV	V_2			Chalcopyrite	$CdSnAs_2$
□		II					VII_2	Distorted defect blende	HgI_2
□			III_2			VI_3		Defect blende	In_2Te_3
	I_2			IV		VI_3		Distorted chalcopyrite	Cu_2SnTe_3
	I_3				V	VI_4		Famatinite	Cu_3SbS_4
		II_2		IV				Fluorite	Mg_2Sn
	I	II			V			?	LiMgSb
□		II_3			V_2			Defect fluorite	Mg_3As_2
	I_3		III		V_2			?	Li_3GaN_2
				IV		VI		Rock salt	PbS
			III		V	VI_2		Wolfsbergite	$AgSbS_2$
□				IV			VII_2	Cadmium iodide	PbI_2

Ternary formulae can be derived from binaries in a similar way, but isoelectronic substitution on the metalloid sites would seem to allow actually existing compounds to be obtained much less frequently. In addition, blende can receive 25% atomic vacancies, so that such vacancies can be made to act as elements with a valency of nil. Table 2.3 gives some examples of the formation of ternary compounds, with or without vacancies. The structures of wolfsbergite and cadmium iodide are very much deformed in relation to that of rock salt, but in an initial approach the bonds are of the same type. The values of E in all these compounds vary considerably. Values of μ_N are usually low, with a very few exceptions (more than 10,000 cm^2/V sec for $CdSnAs_2$).

2.4. Second condition

The second condition is more recent, and is based on a certain interpretation of the concept of ionicity. The passage from one homoatomic molecule similar to hydrogen H_2 to a heteroatomic molecule similar to the molecule of a hydracid HX can be handled by each of the two main methods of theoretical chemistry: the method of molecular orbitals and

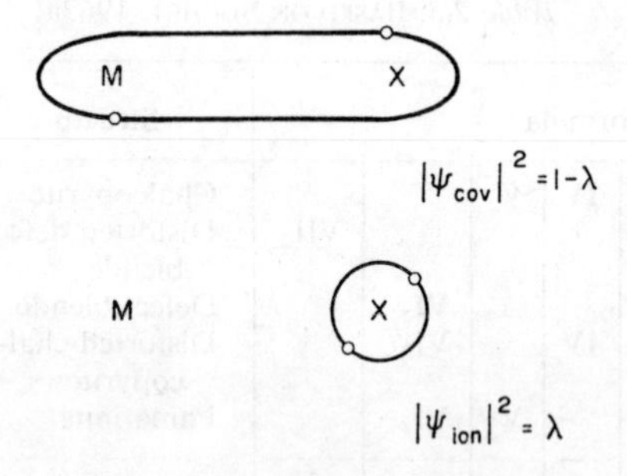

FIG. 2.6. Covalent (top) and ionic (bottom) electronic distributions (based on Suchet, 1961).

the valence bond method. In this second method, which is much closer to chemists' usual concepts, a pair of electrons, known as the "bonding pair", between an atom of metal M and an atom of metalloid X, is described by a common orbital function. We will look for a function expressing a mesomerism between the actual Lewis electronic pair, characteristic of a purely covalent bond, and the electronic doublet of the anion X, characteristic of a purely ionic bond (Fig. 2.6). The probability of existence λ of the electronic doublet may be regarded as a parameter of ionicity. In 1961, we suggested writing, in a simplified form:

$$\psi = (1-\lambda)^{\frac{1}{2}}\varphi_M(1)\varphi_X(2) + \lambda^{\frac{1}{2}}\varphi_X(1)\varphi_X(2),$$

$$\psi = [(1-\lambda)^{\frac{1}{2}}\varphi_M(1) + \lambda^{\frac{1}{2}}\varphi_X(1)]\varphi_X(2).$$

Electron (2) in the electronic pair is then always linked to the X atom, while electron (1) is linked to M if the bond is covalent, or X if it is ionic, and is generally shared between the two.

Tracing an arbitrary boundary between the two atoms, one can accordingly say that the pair of electrons providing the interatomic bond are divided up into $(1-\lambda)$ on M and $(1+\lambda)$ on X. The position of the boundary is determined by the wave function above, which takes a purely covalent state as 0 on the ionicity scale. Pauling, on the other hand, referred to a homopolar state, in other words without charges, accessible from neutral atoms without any transfer of electrons from one atom to another. These two definitions are identical for hydracid HBr, where a purely covalent bond $H^{(s)}Br^{(p^7)}$ does not require any electronic transfer and thus corresponds to a homopolar bond. They differ, however, for most compounds, e.g. blende ZnS, in which the covalent bond $Zn^{(sp^3)}S^{(sp^3)}$ requires the transfer of two electrons to the zinc, thus charging this metal negatively. In general, therefore, the homopolar state, without charges, already corresponds to a certain ionicity λ_0 (called *chemical ionicity*).

In the first condition for semiconduction, the term c was used for the number of electrons contributed by each atom to bonding pairs with its neighbours, namely the number of shared electronic pairs in which it participates. The number of electrons in such pairs is accordingly $2c$, divided up into $c(1-\lambda)$ on M and $c(1+\lambda)$ on X. In the neutral or homopolar state, M possesses n electrons, the very ones it loses when it becomes ionized, and X accordingly possesses the rest, namely $2c-n$ electrons. This distribution must coincide with $c(1-\lambda_0)$ and $c(1+\lambda_0)$,

Table 2.4. BASED ON SUCHET, 1961

sp^3 tetrahedral bond (B3)	$c = 4$	
IV elements	$n = 4$	$\lambda_0 = 0$
III V compounds	$n = 3$	$\lambda_0 = 0.25$
II VI compounds	$n = 2$	$\lambda_0 = 0.50$
I VII compounds	$n = 1$	$\lambda_0 = 0.75$
p^3 trigonal bond (A7 or Bl)	$c = 3$	
V elements	$n = 3$	$\lambda_0 = 0$
II VI or IV VI compounds	$n = 2$	$\lambda_0 = 0.33$
I VII or III VII compounds	$n = 1$	$\lambda_0 = 0.67$

which occurs for

$$\lambda_0 = 1 - n/c.$$

Chemical ionicity depends only on the position of the elements M and X in the Periodic Table of Elements and the type of bond, as is shown in Table 2.4.

The general case of a heteropolar state is obtained from the homopolar state by transferring a certain number q of electrons to an atom X at the expense of the M atoms surrounding it. M then possesses $c(1-\lambda_0)-q$ and X possesses $c(1+\lambda_0)+q$ electrons. This distribution must coincide with $c(1-\lambda)$ and $c(1+\lambda)$, which occurs for

$$\lambda = \lambda_0 + q/c.$$

The ionicity of a molecule or crystal is accordingly obtained by adding to the chemical ionicity λ_0 an *electrostatic ionicity*, a fraction of the charge of the atom affecting each of the shared electronic pairs. Suchet *et al.* (1965) subsequently showed that the ionicity thus defined was practically identical with the result of direct calculation based on a comparison of the bonding energy deduced from thermochemical data with the theoretical energies of the covalent and ionic states.

If one now tries to substitute an antibonding function ψ' for the bonding function considered earlier, one is led to define a parameter λ', such that the nominal distributions of electrons on M and X are reversed. This suggests that the negative values of λ to some extent represents the "metallicity" of the bond, and corresponds to a free electron gas (Suchet and Bailly, 1965). One thus arrives at the second condition of semiconduction, which can be written very simply

$$\lambda > 0.$$

Let us calculate, for instance, the ionicities of indium compounds in the B3 structure:

InN	InP	InAs	InSb	InBi
0.65	0.61	0.42	0.24	−0.11

InBi crystallizes in a tetragonal structure, but the arrangement of atoms and their distances justify comparing this compound with the preceding ones in the B3 structure. Its metallic character is quite consistent with the negative sign found for λ if its ionicity is calculated.

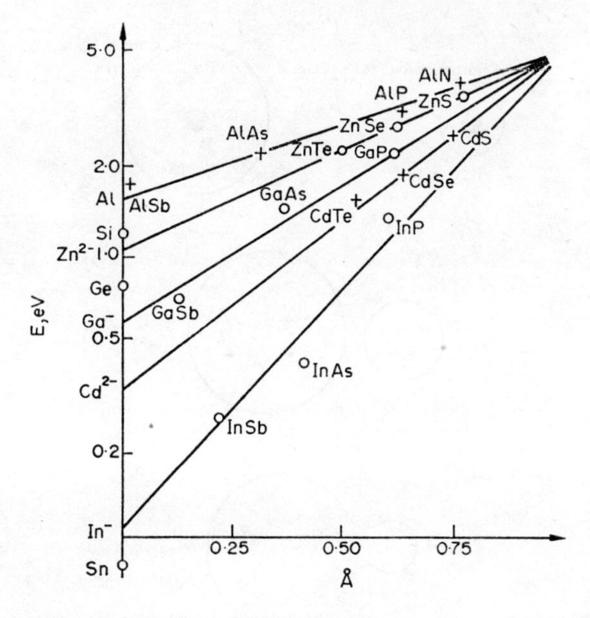

FIG. 2.7. Variation in logarithm of E of some binary compounds with B3 structure with relation to λ (based on Suchet, 1962b).

A number of physical measurements vary simply in relation to λ: e.g., the parameter E of equiatomic binary compounds in the B3 structure (Fig. 2.7).

2.5. Effective atomic charge

The charge q considered in the previous section lies between the ionic charge n created when a neutral atom is ionized by the transfer of n electrons from M to X, and the charge m which appears in the covalent state after the transfer of m electrons from the neutral atom X to the neutral atom M. Positive or negative, the algebraic value of this *effective atomic charge* thus complies with the double inequality

$$-m < q < n.$$

In 1961, we suggested the following procedure to evaluate q. Take two ions, possessing Z and Z' electrons respectively, the nuclei of which carry charges $(Z+n)$ and $(Z'-n)$ (Fig. 2.8a). When the ions approach

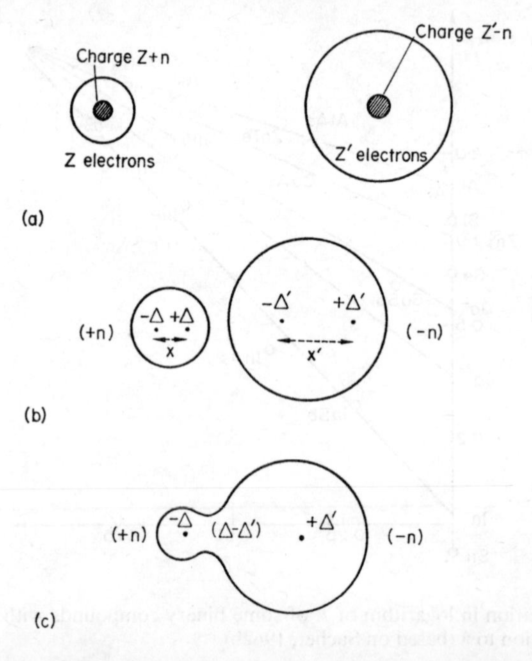

FIG. 2.8. Schema of the mutual polarization of two ions, which partially leads back to the covalent electron configuration: (*a*) without interaction, (*b*) with weak interaction, (*c*) with overlap (based on Suchet and Bailly, 1965).

each other, the resulting charges $+n$ and $-n$ create electric fields which produce opposing forces on the partner's nucleus and electron cloud. This results in dipolar moments Δx and $\Delta' x'$ in each ion (Fig. 2.8*b*). When the electron clouds interpenetrate, reflecting development towards a covalent bond, the neighbouring charges Δ and $-\Delta'$ neutralize each other partially, and leave only the residual charge $(\Delta - \Delta')$ (Fig. 2.8*c*). If this residue is shared equally between the two ions, to simplify matters, one finds

$$q = n - (\Delta + \Delta')/2.$$

We then introduced the ionic radii r and r', in order to make n/r^2 and $-n/r'^2$ comparable to average fields, and assumed that Δx and $\Delta' x'$ are proportional to nZ/r'^2 and nZ'/r^2. But x and x' vary in inverse ratio to r and r', so that

$$q \sim n[1-(Z/r'+Z'/r)C^{t}].$$

Assessment of the constant at 0.01185 gives zero charges on the elements germanium and tin. Batsanov, who also introduced two chemical and electrostatic components of ionicity, considers only deformation of the electron cloud of the anion, which increases in direct proportion to the electronegativity χ of the cation, and in inverse proportion to the ionization potential I of the anion. The valency electrons are divided up into $\chi(I+\chi)$ on the cation and $I(I+\chi)$ on the anion, and the difference between these two numbers indicates the electronic transfer from the anion towards the cation. Batsanov arrives at an expression similar to the preceding one:

$$q \sim n[1-(\chi-I)/(I+\chi)].$$

The effective charge thus defined is a *static* charge, closely connected to the elementary dipole of the crystal (Suchet, 1973). This will intervene in various properties of the lattice, and notably to limit carrier mobility, when the other limiting factors can be ignored. This is why the zero values found for crystals with high mobility, such as InSb and HgTe, from the start gave a certain credibility to q charges, despite the vagueness of their evaluation. In addition, the diffusion of charged interstitial impurities will introduce couplings with metal or metalloid vacancies, depending on the sign of q. The diffusion of Li^{+} ions in III V compounds thus results in a coupling with a negative V vacancy (in the case of a positively charged V atom) in AlSb and GaSb, and with a negative III vacancy (in the case of a positively charged III atom) in GaAs. Finally, in the case of low-ionicity compounds, replacement of the ionic charge by the effective charge allows the electrostatic theory of ligands to be extended, and applied to qualitative prediction of certain physical effects.

Other phenomena can evoke the presence of an effective charge on atoms: the dielectric properties of crystals (cf. Chapter 5). If one tries to connect dielectric constants at low and high frequency and infrared dispersion frequencies with microscopic data, one is led to attribute the existence of dipolar moments to the deformation and movements of atoms. At high frequencies, such as those of infrared waves, movements introduce *dynamic* effective atomic charges, referred to by Szigeti by the symbol e^{*}.

The relation between the two types of charges has been explained (Suchet and Bailly, 1965; Suchet, 1973) by assuming that the former depend on the relative movement x of M and X atoms:

$$q(x) = q(0) + x\mathrm{d}q/\mathrm{d}x.$$

The dipolar moment P can be written in two different ways: $\nu e^* x$, where ν refers to the number of MX "molecules" per unit of volume, and $\nu x[q(0)+R\mathrm{d}q/\mathrm{d}x]$, where the first term expresses the relative movement of two atoms with charges $\pm q(0)$, and the second the transfer of the charge $\mathrm{d}q/\mathrm{d}x$ over their mutual distance apart R. One has

$$e^* = q(0) + R\mathrm{d}q/\mathrm{d}x,$$

and the sensitivity of the static charge to small relative movements of the atoms can be seen. The same end would be attained if, the atomic nuclei being fixed, the electron clouds would be submitted to a periodical perturbation.†

Table 2.5. BASED ON SUCHET AND BAILLY, 1965

Formula	Static charge	Dynamic charge
GaP	1.50	0.58
InAs	0.68	0.56
GaAs	0.49	0.51
InSb	−0.04	0.42
GaSb	−0.50	0.33
AlSb	−0.98	0.53

Table 2.5 shows the difference in value generally existing between the two types of charges, for some III V compounds. Despite the imprecision of the evaluation of static charges, these would appear to vary much more from one compound to another than dynamic charges. For several years, attempts have been made to connect such charges to radiocrystallographic data: displacement of satellite lines $K\alpha_1$ and $K\alpha_2$, or a map of the distribution of electron density followed by graphic integration (Fig. 2.9). Interpretation of such work is not unanimously accepted, however.

† In a recent work, Phillips defines an "ionicity" from the dielectric constant at optical frequencies. Such a definition implicitly refers to a *dynamic* effective charge (Suchet, 1974).

FIG. 2.9. Electron density levels in AlSb crystals (based on Sirota and Gololobov, 1964).

References

KREBS, H. (1968) *Grundzüge der anorganischen Kristallchemie*, Enke, Stuttgart (English translation, McGraw-Hill, Maidenhead).

LINDBERG, O. (1952) *Proc. IRE* **40,** 1414.

MOOSER, E. and PEARSON, W. B. (1956) *Report of a Meeting on Semiconductors, Rugby 1956*, Physical Society, London.

NEUBERGER, M. (1971) *III V Semiconducting Compounds*, IFI/Plenum, New York.

PAULING, L. (1960) *The Nature of the Chemical Bond*, Cornell, Ithaca, NY.

PHILLIPS, J. C. (1970) *Rev. Mod. Phys.* **42,** 317.

SHOCKLEY, W. (1950) *Electrons and Holes in Semiconductors*, van Nostrand, New York.

SIROTA, N. N. and GOLOLOBOV, E. M. (1964) *Dokl. Akad. Nauk SSSR* **156,** 1075.

SUCHET, J. P. (1961) *J. Phys. Chem. Solids* **21,** 156.

SUCHET, J. P. (1962a) *Chimie physique des semiconducteurs*, Dunod, Paris (English translation, van Nostrand, London, 1965).

SUCHET, J. P. (1962b) *C.R. Acad. Sci. Paris* **255,** 1444.

SUCHET, J. P. and BAILLY, F. (1965) *Annls Chim. Paris* **10,** 517.

SUCHET, J. P. (1973) *Bull. Soc. Chim. Fr.* 922.

SUCHET, J. P. (1974) *C.R. Acad. Sci. Paris* **278 C,** 1361.

SZIGETI, B. (1949) *Trans. Faraday Soc.* **45,** 155.

Chapter 3

Magnetic Semiconductors

3.1. Roles of s, p, and d electrons

We shall now turn to compounds containing ordered or disordered atomic magnetic moments, showing paramagnetism of ferro-, ferri-, or antiferromagnetism. They include a metal T, belonging to the transition metals or metals like copper or silver, in which the d shell can be incomplete in certain valency states, or even possibly a lanthanide L or actinide A. They also include a metalloid X, belonging to the elements in columns IV, V, VI and VII of the Periodic Table of Elements. In contrast to what happens in alloys, the cohesion of the crystal of such a compound mainly results from the cohesion of the T–X bonds, and the respective roles of s and p electrons on the one hand, and d electrons on the other, can be dealt with more or less separately.

Let us look again at the types of interatomic bonds described earlier. If the T–X bond is purely ionic, the s and p electrons that the T atoms possessed in the neutral state have completed the octet of X. They therefore entirely fill a system of energy levels on anions, and return to cations cannot occur without additional external energy. One can obtain an idea of the energy difference E separating these two systems by referring to alkaline fluorides, where it is approximately 8–9 eV. If the T–X bond is purely covalent, it can be isolated arbitrarily from the rest of the crystal, and described like the bond between the two hydrogen atoms in the hydrogen molecule. It will be remembered that the energy of the molecule is calculated in relation to the distance the nuclei are apart. At the equilibrium distance, one still finds an energy difference E of approximately 9 eV between the two possible states. Another description, not so well adapted to the present case, consists of

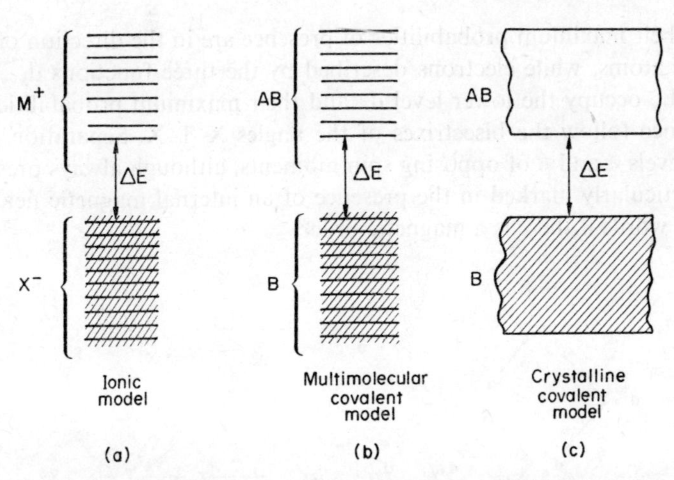

FIG. 3.1. Theoretical models for electrical conduction in crystal inorganic compounds (based on Suchet, 1971).

classifying the electrons of the whole crystal on an energy basis. Semiconductors, then, possess two separate systems—valency electrons and conduction electrons. As shown in Fig. 3.1, examination of these three situations—ionic model, multimolecular covalent model, and crystalline covalent model—ultimately produces a single representation. The s and p electrons in the T–X bonds occupy two systems of levels or two energy "bands" in the crystal, the lower one of which is completely filled at absolute zero, while the upper one is empty. To describe them, we shall use the terms proposed by Heitler and London, *bonding* and *antibonding*.

Let us now consider the role of d electrons. This is rather complex, and to simplify matters we shall confine ourselves to the simple case in which the metal T is in the centre of an octahedron, more or less deformed, of six X neighbours with which it is directly linked. In this case, an initial approximation suggests that the ten d electrons are divided into two groups of energies: six occupy a lower level $d\epsilon$ and four an upper level $d\gamma$. Two factors contribute to this splitting: an electrostatic effect involving the charge borne by each ligand X, and the covalent effects connected with considerations of symmetry and orientation in the d atomic orbital functions. As is shown in Fig. 3.2, electrons described by the two functions $d_{x^2-y^2}$ and d_{z^2} occupy the upper level $d\gamma$

and their maximum probabilities of presence are in the direction of the six X atoms, while electrons described by the three functions d_{xy}, d_{yz}, and d_{zx} occupy the lower level $d\epsilon$ and their maximum probabilities of presence follow the bisectrixes of the angles X–T–X. Separation into sub-levels α and β of opposing spin moments, although always present, is particularly marked in the presence of an internal magnetic field, in other words if there is a magnetic order.

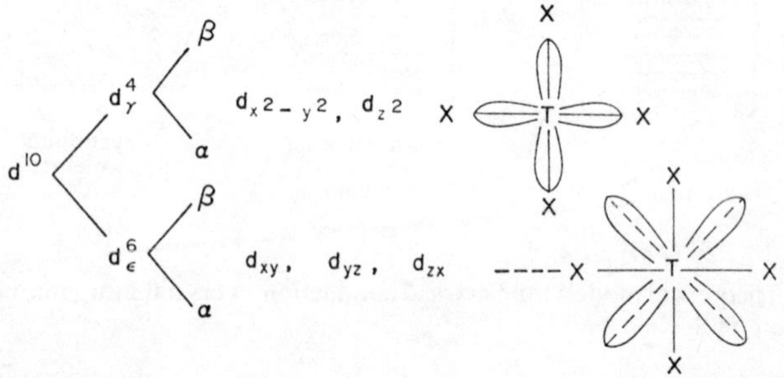

FIG. 3.2. Simplified schema of the splitting of the d electron level of a T atom in an octahedral environment of X atoms.

Finally, after defining the position of the systems of bonding and antibonding levels of s and p electrons, and then of d electron levels and sub-levels, one has ultimately to compare their respective situations. This is where the difficulty lies. If the average atomic weight of the elements T and X is low, the T–X bond is relatively ionic, the value of E is high, and the $d\epsilon$ and $d\gamma$ levels are situated between the bonding and antibonding systems, as shown in Fig. 3.3. This layout applies, for example, to iron, cobalt, and nickel monoxides crystallizing in the structure of rock salt. If, on the other hand, the average atomic weight of the elements T and X is high, the T–X bond is relatively covalent and E is lower. Interactions between $d\epsilon$ and $d\gamma$ levels, on the one hand, and the systems of s and p levels on the other, are then possible. Finally, if the bond is strongly covalent, the sign of the effective atomic charges may be inverted, the T atom being negatively charged. It may be assumed that in this case the order of the $d\epsilon$ and $d\gamma$ levels would be inverted. This could be the case for some antimonides crystallizing in the structure of NiAs, such as CrSb.

It has been seen that the maximum probabilities of presence of dγ atomic orbital functions, like p functions, lie in the directions of the six neighbouring metalloids. They can therefore participate in the T–X bond if the energy of the dγ level is close to that of the bonding system. The hybrid orbital function dγ^2p has the same symmetry, notably, as all three p functions, and could, like it, provide a bond by resonance between two groups of three neighbours. This, however, would lead to

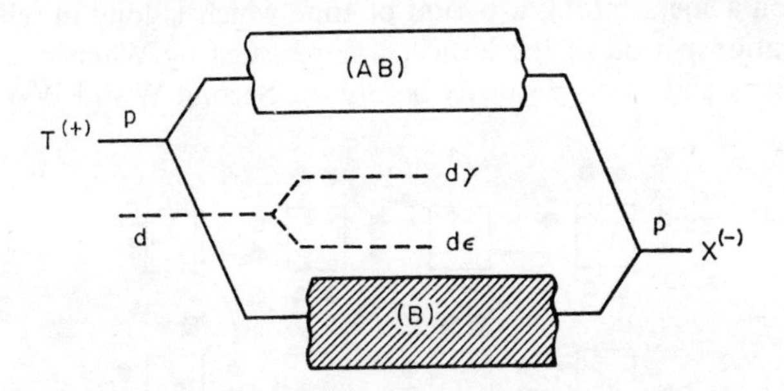

FIG. 3.3. Simplified energy diagram of the bonding and antibonding bands and of the dϵ and dγ sub-levels in the general case (based on Suchet, 1969).

the appearance of surplus electrons, which would occupy the antibonding system, so that such hybridization would result in a metallic character. It is not impossible that this situation may exist in compounds of NiAs structure, such as CoS or FeP. Furthermore, the orbital function dγ^2sp^3 also has the same symmetry as all p functions, and could therefore provide the bond with six neighbours. The atomic ratio X/T must be equal to 2, however. This is the case for rutile TiO_2 and pyrite FeS_2, both semiconductors.

So far we have mentioned only T–X bonds. The existence of direct bonds between T atoms, with compensation of spin moments and two d electrons in a Lewis pair, is also possible if the atoms are close enough together. Goodenough (1963) has explained in this way the magnetic and electric properties of oxides such as V_2O_3 and VO_2 at low temperatures, and of compounds such as $CoAs_2$ in the structure of arsenopyrite.

Finally, we have indicated briefly that separation of each of the levels dϵ and dγ into sub-levels α and β, with opposing spin moments, is very accentuated in the presence of a magnetic order. It is clear that this modification can result in certain interactions between these sub-levels and the systems of bonding and antibonding levels. We shall return to this in § 4.1 in connection with certain metal–semiconductor transitions.

3.2. d transfers and chemical bond

The existence of multiple valencies in transition metal compounds is well known, and a semiconduction mechanism, in which a d electron passes, on a metal atom, a period of time which is long in relation to the vibration period of the lattice, was revealed by Wagner, and then by De Boer and Verwey shortly before the Second World War. Let us

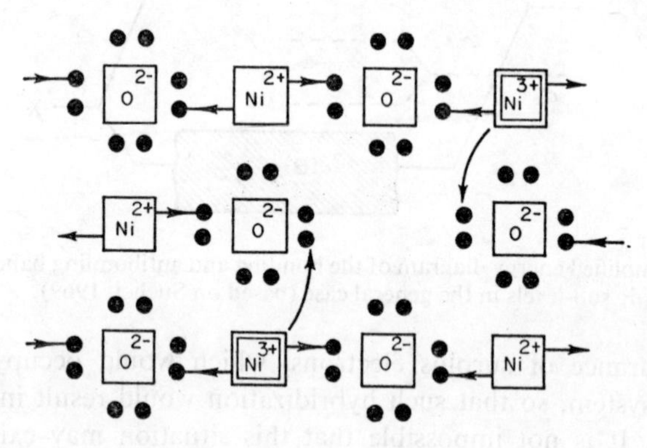

FIG. 3.4. Effect of a nickel vacancy in NiO crystals.

consider the example of the oxide NiO, which crystallizes in the structure of rock salt. Figure 3.4 shows the lattice of this oxide with a nickel vacancy. The two electrons needed to complete the octet of neighbouring oxygens are supplied by two of the nickel atoms, which lose a d electron and thus change to valency III. But these two nickel atoms are taken at random among those surrounding the vacancy. Why are they selected and not others? There is no reason, and one can easily conceive that the hop of a d electron from a divalent nickel (which thereupon becomes trivalent) to a trivalent nickel (which thereupon becomes divalent) is relatively easy. Figure 3.5 shows this *transfer* on one level, in diagrammatical form, the transferred electron ultimately retaining the same energy difference in relation to electrons in lower shells.

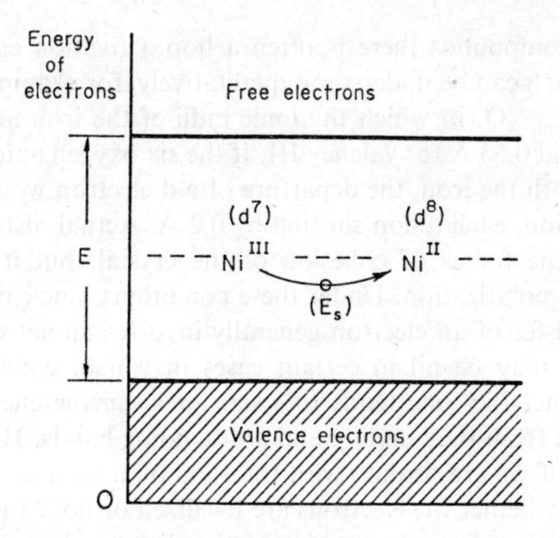

FIG. 3.5. Electron transfer on a d level. Case of defect NiO.

The mechanism by which mixed valencies appear in oxides therefore rests on a *valency induction*, which can be caused by a lattice defect, vacancy or interstitial, or a substitution impurity (De Boer and Verwey). It is fairly easy to predict the number of metal atoms changing valency and write the complete ionic formula, if the condition of electrical neutrality is used. For instance, in the case of a metal vacancy

$$\square_x T^{n+}_{1-x} O^{2-},$$

the average ionic charge n carried by a T atom is given by the condition

$$n = 2/(1-x)$$

and n cannot be a whole multiple of the elementary charge unless $x = 0$ $(n = 2)$, $x = \frac{1}{3}$ $(n = 3)$, and $x = \frac{1}{2}$ $(n = 4)$. For any other value of x, the low ionization energy of the d electrons points to the appearance of two different valencies

$$\square_x T^{n+}_y T^{n'+}_z O^{2-},$$

where the indices y and z are easily evaluated in relation to x by once again writing the condition of neutrality, then the invariance of the total number of atoms. If these atoms occupy equivalent sites in the crystal, transfers will be possible.

In ionic compounds there is often a hop activation energy E_S, the origin of which can be understood qualitatively, for example in the case of wüstite $Fe_{1-x}O$, in which the ionic radii of the iron are 0.75 Å for valency II and 0.53 Å for valency III. If the six oxygen anions remained in contact with the iron, the departure of a d electron would produce a local distortion, each anion shifting by 0.2 Å. Actual distortion is less, because of the forces of cohesion of the crystal, but it nevertheless causes slight polarization. Under these conditions, one can understand that the transfer of an electron generally involves an activation energy E_S, but this may be nil in certain cases in which, conduction being apparently metallic, one tends to speak of a *narrow* energy band, to distinguish it from the bonding and antibonding bands. How are we to know, then, if we are dealing with transfers on a level or in a band, in other words whether the electrons are localized or not? In 1960, Morin noticed that this depends mainly on the distance between second T neighbours. The mathematical theory was presented the following year by Mott, but it was Goodenough (1963) who provided practical rules allowing the critical distance R_c, below which delocalization occurs, to be assessed in every structure.

If a covalent compound is involved, we have shown (Suchet, 1967) that the number of metal atoms changing valency can also be predicted. X metal vacancies correspond to an equal number of X atoms with a non-bonding electronic pair. Take, for example, a rock-salt structure in which three pairs of p electrons resonate between six neighbours. If the metal has the electronic formula $d^k s^2$, the crystal-formation reaction is expressed

$$(1-x)T^{d^k s^2}+X^{s^2 p^4}\longrightarrow x\square X^{s^2 p^6}+(1-x)T^{d^{k'} s^{0(p^3)}}X^{s^{2(p^3)}}.$$

Instead of the electrical neutrality condition, we shall here use the condition of invariance in the total number of electrons (d+s+p)

$$k' = k-2x/(1-x)$$

which will be an integer if $x = 0$ ($k' = k$), $x = \frac{1}{3}$ ($k' = k-1$), or $x = \frac{1}{2}$ ($k' = k-2$). For any other value, as for the ionic crystal, two different valencies appear, and one finds the same indices y and z. This confirms, what has not always been clearly illustrated, that laws governing the appearance of a mixed valency are completely independent of the more or less ionic nature of bonds. However, there must be some reservation in extending them to covalent compounds. To begin with, there cannot be any transfer of electrons between two T atoms with neighbouring valencies unless their d orbital functions coincide in direction and overlap sufficiently. Secondly, the hop activation energy E_S is much lower here, since distortion and local polarization disappear. If it is eliminated, transfer will occur freely on a widened energy level, and conduction will be apparently metallic. This brings us back to the case of the narrow energy band already encountered in connection with oxides, but the concept of critical distance is less evident here.

The probability of transfers of d electrons between neighbouring metal atoms can be reasonably well predicted, on the basis of the rate of filling of the d levels, the degree of coincidence of the directions of the corresponding orbitals and the extent to which they overlap. The first of these is the most important: transfer cannot occur unless one of the d sub-levels is partly filled. Transfer cannot take place in an empty level or in a full level. It was seen previously that, in an octahedral environment (as in rock salt, rutile, corundum, etc.), six d electrons could occupy a lower sub-level $d\epsilon$ and four an upper sub-level $d\gamma$. In fact, the successive electronic formulae for progressive filling of the d layer are given in Table 3.1. It will be seen that the sub-levels $d\epsilon\alpha$, and then $d\gamma\alpha$, where the spin moment is in the same direction, fill up before sub-levels $d\epsilon\beta$, then $d\gamma\beta$, where the moment of the electrons is opposed to that of the α sub-levels. Whenever a formula contains only empty or completely full sub-levels, it cannot lend itself to transfers, and has been

Table 3.1. BASED ON SUCHET, 1973 (OCTAHEDRAL ENVIRONMENT)

d^1	$d\epsilon^1\alpha$
d^2	$d\epsilon^2\alpha$
d^3	~~$d\epsilon^3\alpha$~~
d^4	$d\epsilon^3\alpha d\gamma^1\alpha$
d^5	~~$d\epsilon^3\alpha d\gamma^2\alpha$~~
d^6	$d\epsilon^3\alpha d\gamma^2\alpha d\epsilon^1\beta$
d^7	$d\epsilon^3\alpha d\gamma^2\alpha d\epsilon^2\beta$
d^8	~~$d\epsilon^3\alpha d\gamma^2\alpha d\epsilon^3\beta$~~
d^9	$d\epsilon^3\alpha d\gamma^2\alpha d\epsilon^3\beta d\gamma^1\beta$
d^{10}	~~$d\epsilon^3\alpha d\gamma^2\alpha d\epsilon^3\beta d\gamma^2\beta$~~

struck out in the table. In such cases, as in conventional semiconductors, one finds the mechanism of semiconduction by the exciting of an electron to a higher energy through the forbidden region E.

3.3. Simple binary compounds

It is the B1 structure of rock salt that is usually considered the simplest. Let us begin with it and first examine the case of *interstitial* compounds in which the distance between T atoms and the physical properties are often close to those of the pure metal. In this structure they mainly comprise some carbides and nitrides, in which the first level $d\epsilon\alpha$ is never completely full (TiC: $d\epsilon^0\alpha$, VC and TiN: $d\epsilon^1\alpha$), and the oxide TiO ($d\epsilon^2\alpha$) can be added. Let us take the example of TiN. Figure 3.6 shows that if the p and $d\gamma$ orbitals of the central titanium atom are directed towards the nitrogen atoms, the $d\epsilon$ orbitals are directed along the bisectrixes of the T–X–T angles, in other words exactly towards the twelve T second neighbours. Overlapping is excel-

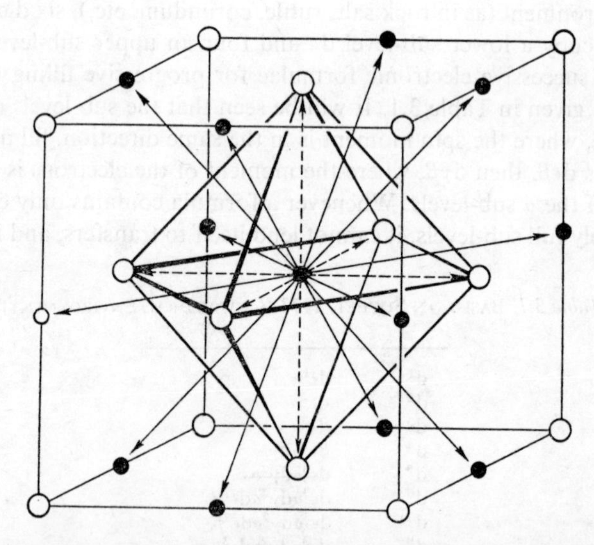

FIG. 3.6. Directions of maximum electron densities of the $d\epsilon$ atomic orbitals (usual arrows) and of the p or $d\gamma$ atomic orbitals (dotted arrows) on the T atoms of a TX compound with the B_1 structure (based on Kjekshus and Pearson, 1964).

lent, so that a high density of transfers to a widened d$\epsilon\alpha$ level can be predicted. In fact, an apparent metallic behaviour is observed, in other words the activation energy E_S is nil. A theoretical model to explain the behaviour of such compounds has been proposed by Bilz. It comprises a system of T–X bonding levels or bonding band, entirely filled, a T–X antibonding system or antibonding band, completely empty, and then a band corresponding to the energies of the d electrons capable of moving freely, the particular feature of which is that it slightly overlaps the other two bands. At the lower part there is a band corresponding to the s electrons (non-bonding) of the X atoms. This model explains the low density of transfers in TiC (d$\epsilon^0\alpha$), where the mechanism of semiconduction by exciting can be observed, and the behaviour of the solid solutions TiC_xN_{1-x} and $Ti_xV_{1-x}C$.

For normal compounds, the B1 structure illustrates the typical case of *superexchange* magnetic interaction. This mechanism can occur in the pure state only in insulating crystals. It results from the existence of partly covalent T–X bonds, which can resonate on each side of the metalloid, and the value of the interaction accordingly depends on the angle between these two successive bonds. It also depends on the extent of the covalent aspect, and varies correspondingly to this, in accordance with a simple law illustrated in Fig. 3.7, which includes in particular the series MnO, MnS, MnSe, all three with B1 structure. When the metal atom is much smaller in dimensions than the X atom (the case of T transition metals), the pure T–T superexchange interaction results in antiferromagnetism. This is the case for the series MnO, MnS, and MnSe, to which the configuration d$\epsilon^3\alpha$d$\gamma^2\alpha$ confers remarkable stability and an absence of transfers. When this is no longer the case, e.g. with lanthanides L, one also has to take account of a *direct* interaction between the moments of the L atoms. This interaction predominates in EuO, EuS, and EuSe, where it results in ferromagnetism, but becomes less with large tellurium atoms, and antiferromagnetism resulting from the superexchange reappears in EuTe.

The large number of lanthanide compounds crystallizing in the B1 structure requires some comment. The deep f level is particularly stable when half-filled with seven electrons, so that it is convenient to distinguish two sub-levels fα and fβ with opposing spin moments. There cannot be any f transfers, since overlapping of the electronic orbitals in

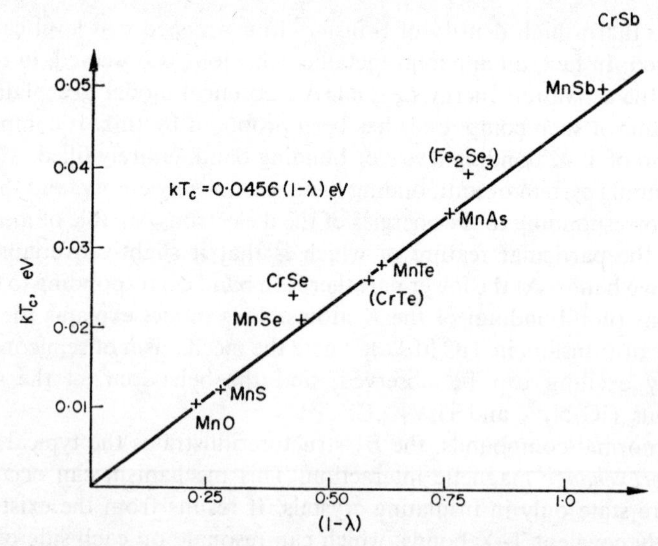

FIG. 3.7. Variation in kT_C magnetic interaction energy of the bonding pair T–X in relation to the $(1-\lambda)$ probability of a purely covalent electron distribution (based on Suchet, 1969, 1971).

this level is not sufficient, but f–d transitions often fill the d level, giving the compound a high density of transfers and metallic character. This is the case, in particular, of divalent compounds, which are not very stable, except for Sm (f^6), Eu (f^7), and Yb (f^{14}). The same probably applies to actinides. In stable compounds, at average temperatures, a low activation energy E_S is often observed, corresponding to d transfers, and then at higher temperatures the activation energy E, corresponding to the f–d exciting of the intrinsic semiconductor. Even in such compounds, what is more, filling of the d level may be caused, for instance, by partial replacement of an Eu (f^7) atom by La (f^1) or Gd (f^8) atoms, in which valency II is not stable. Resistivity thus drops very quickly in $Eu_{1-x}Gd_xSe$ solid solutions when the rate of x substitution increases. A new phenomenon then appears: interaction between the magnetic order of the europium atom moments and the orientation of the spin moments of electrons during transfer. This is the Ruderman–Kittel–Yosida interaction.

The slightly less straightforward C4 structure of rutile might also be mentioned, for comparison. In this structure, the T atom is still in the centre of an octahedron of six neighbouring metalloids, but it is connected to them by six hybrid $d\gamma^2sp^3$ orbitals. As for the metalloid atoms, generally oxygen, they are connected to the three neighbouring T atoms by three hybrid sp^2 orbitals, forming an angle of only 120° with one another, which is not very favourable to magnetic superexchange interactions. The extensive overlapping of $d\epsilon$ orbitals results in a high density of transfers in a narrow band, the appearance of which in reduced TiO_2 is well known, and which is also found in VO_2 ($d\epsilon^1\alpha$) and CrO_2 ($d\epsilon^2\alpha$). But the completely full sub-level of βMnO_2 ($d\epsilon^3\alpha$) prevents transfers and allows one to observe semiconduction by exciting, while the magnetic order takes on the form of a helicoidal antiferromagnetism. The case of CrO_2, however, requires some comment. According to Goodenough's rule, the distance between chromium atoms is greater than the critical delocalization distance, and a transfer activation energy should therefore be observed: this is not the case. It almost occurs, however, since optical study of $Cr_{1-x}Mn_xO_2$ solid solutions reveals the appearance of an optical absorption edge once $x = 0.1$. The manganese magnetic moments are opposed to those of chromium, and the magnetic structure of the solutions is not colinear.

3.4. Compounds with the NiAs structure

Many binary compounds of transition metals crystallize in the $B8_1$ structure of nickel arsenide. This structure, which is very common, is nevertheless rather unfamiliar. The X atoms form misshapen octahedrons round the T atoms, but the X–T–X angles remain close to 180°. The T–X bonds are formed mainly by p orbital functions, so that this structure resembles the B1 structure of rock salt. This results from interpretation of the infrared spectra of chromium compounds, and of the electrical and magnetic properties, which would not be the same in the case of intervention of hybrid $d\gamma^2sp^3$ orbitals. Planes perpendicular to the c axis, consequently noted (001), succeed one another along this axis, and contain T atoms and X atoms alternately. This arrangement is important, since the predominant magnetic interactions are those of superexchange (although the angle T–X–T here drops to approximately 130°) and the T moments of one (001) plane will be

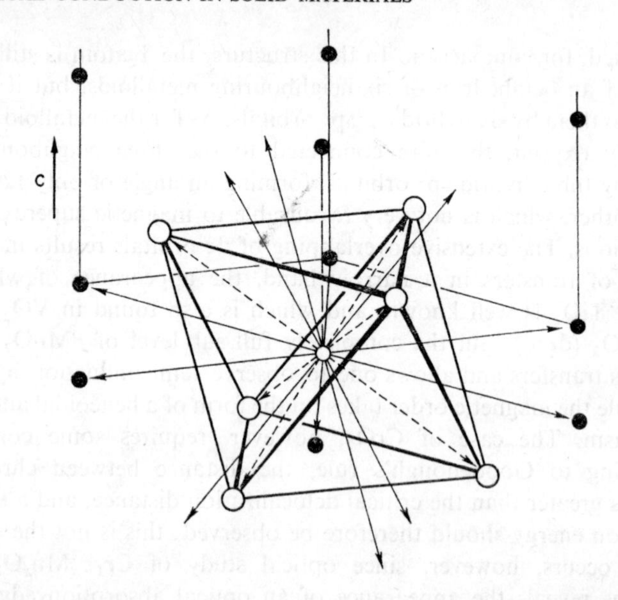

FIG. 3.8. Directions of maximum electron densities of the dϵ atomic orbitals (usual arrows) and of the p or dγ atomic orbitals (dotted arrows) on the T atoms of a TX compound with the $B8_1$ structure (based on Kjekshus and Pearson, 1964).

orientated antiparallel to the T moments of the following (001) plane. This means that there will quite often be antiferromagnetism from plane to plane.

In addition, the $B8_1$ structure, in contrast to the B1 structure, shows high anisotropy, as illustrated in Fig. 3.8. The black dots represent T atoms and the white circles X atoms. The six T atoms at the bottom belong to the same (001) plane, and the directions of their dϵ orbitals, indicated by uninterrupted arrows, coincide with those of their second neighbours in this plane. With their second neighbours in the plane above, on the other hand, they form an angle well below 180°, which is not very favourable to transfers in the direction of the *c* axis. Although the distance between T second neighbours along the *c* axis is generally shorter than in an (001) plane, transfers would therefore appear to be easier in the second case (Suchet, 1971). This has been verified by experiment in two very different cases: the case of a cobalt telluride with metallic behaviour, in which transfers took place in a narrow d band, and the case of an iron sulphide with semiconductor behaviour in which transfers comprise an energy E_S above zero. In both cases, however, defect compounds are involved, and this requires some explanation.

While there is often a slight interstitial excess of transition metal in T V compounds, there is a shortage in T VI compounds, which are frequently represented by a formula $\square_x T_{1-x} X$. The variable range of non-stoichiometric compositions often comprises monoclinic over-structures for around $x = \frac{1}{8}$ and ceases for around $x = \frac{1}{4}$ or $\frac{1}{3}$. With TiSe, TiTe, and NiTe, however, $x = \frac{1}{2}$ can be reached, with one in every two T atoms completely disappearing. Without encountering any phase change, one thus has the C6 structure of cadmium hydride, which is a layer structure. In defect NiAs structures, T metal vacancies are usually ordered, and affect only one T atom plan in two. The structure consists in turn, along the c axis, of a T plane with vacancies, an X plane, a full T plane, an X plane, and so on.

Figure 3.9 shows four different cases, depending on whether the valency induction acts in the same plane as vacancies are located or in the full T plane, and depending on whether the vacancy ratio results in valencies II and III coexisting ($0 < x < \frac{1}{3}$) or valencies III and IV ($\frac{1}{3} < x < \frac{1}{2}$). In each of these cases a simple calculation involving either the condition of neutrality of the crystal, using the ionic layout, or the condition of invariance of the full number of d, s, and p electrons, using the covalent layout, will show the proportion of each valency, and consequently the density of transfers in the (001) planes with mixed valencies. If it is assumed that conduction along the c axis can be ignored, at least to begin with, transfer-density variations are obtained in the different cases, disappearing for certain values of x. These zero minima correspond either to the eventuality of the crystal containing only one valency ($x = 0$ for valency II, $x = \frac{1}{3}$ for valency III, $x = \frac{1}{2}$ for valency IV), or to the eventuality of all the atoms of one valency being situated in planes with vacancies, and all the atoms of the other in full planes, which naturally involves different values for x, in the cases illustrated in Fig. 3.9.

These problems of transfers in the structure NiAs have been considered in detail because the values of x corresponding to zero or almost zero transfer minima in the (001) planes appear to be observable not

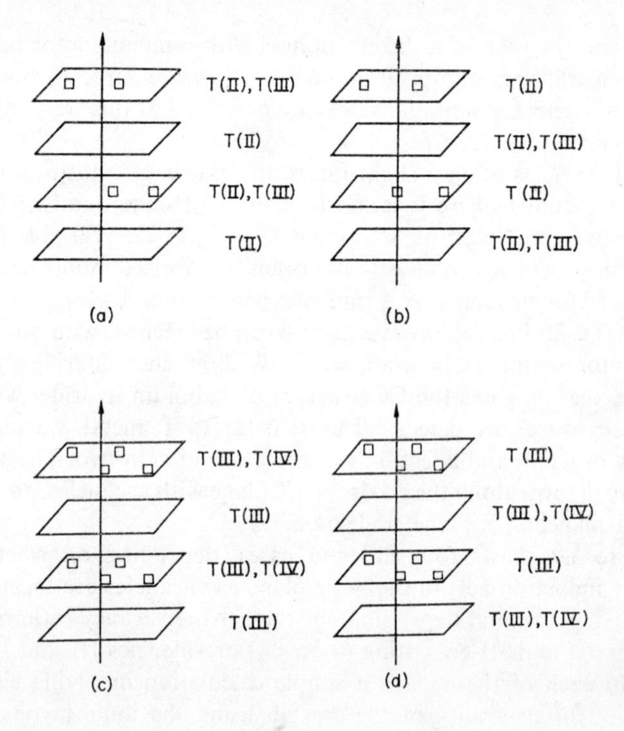

FIG. 3.9. Various cases of the presence of multiple valencies in planes containing vacancies (*a*, *c*) or full planes (*b*, *d*). $0<x<\frac{1}{3}$ in (*a*) and (*b*), $\frac{1}{3}<x<\frac{1}{2}$ in (*c*) and (*d*) (based on Suchet, 1967, 1971).

only in semiconductor compounds, in which a small density of transfers takes place in a d sub-level, but also in compounds whose apparent metallic behaviour indicates the existence of transfers in a narrow energy band. Valency induction laws cannot apply to delocalized electrons, so that we are forced to recognize that an electron moving in a narrow band is partly localized (Suchet, 1967). This does not imply any real contradiction with physical theories anyway.

Let us now return to the electrical conduction anisotropy mentioned earlier. Figure 3.10 shows, on natural logarithmic scale, the conductivity of a monocrystal of $Fe_{1-x}S$, where x has a very low value. Two measurements have been taken in terms of temperature, perpendicular

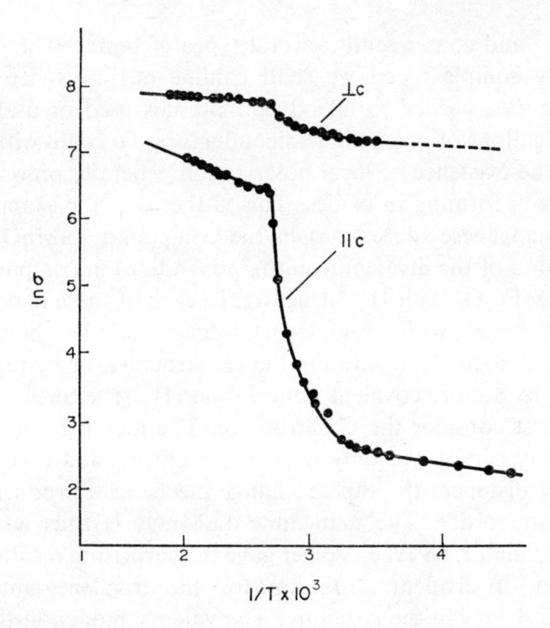

FIG. 3.10. Logarithmic variation of the conductivity of an $Fe_{1-x}S$ monocrystal in relation to the reciprocal of the absolute temperature in the two main directions (based on Kamigaichi *et al.*, 1956).

and parallel to the *c* axis. At low temperatures, on the right-hand side of the figure, conductivity in the (001) plane is much higher than along the *c* axis. But conductivity along the axis increases suddenly as one approaches the temperature at which vacancies become disordered, and the difference thereupon increases considerably. A little later, other scientists discovered a similar break for the Hall coefficient, and proved that it was due not to a variation in the number of carriers but to a variation in the value of their mobility. The mechanism by transfers, which is more sensitive to disorder in the structure, has recently been confirmed by Frolich.

3.5. Ternary compounds

The G5 structure of perovskite and $H1_1$ structure of spinel correspond to ternary compounds involving two types of metals occupying different sites, and consequently several types of bonds. The situation is accordingly complex, and we shall confine ourselves here to a few simple remarks, needed to introduce materials used or usable for the main applications of magnetic semiconductors. To begin with, one may note that the existence of local polarization generally prevents narrow d bands from forming in oxides. This is the case, for example, of the trivalent manganese ($d\epsilon^3\alpha d\gamma^1\alpha$) in the compound $LaMnO_3$ with G5 structure, and of the divalent iron ($d\epsilon^3\alpha d\gamma^2\alpha d\epsilon^1\beta$) in compound Fe_3O_4 or ferrites MFe_2O_4 with $H1_1$ structure. In each of these two structures, however, materials with a high transfer density can be obtained, either by causing a valency induction in the G5 structure, or by replacing the T–O bond by a more covalent bond in the $H1_1$ structure.

Let us first consider the G5 structure. The metal in the octahedral site generally carries the only magnetic moment, and except at large interatomic distances the superexchange mechanism produces an antiferromagnetic order. The manganite $LaMnO_3$ is thus an antiferromagnetic insulator. In 1950, Volger gave the surprising results of partial substitution of divalent strontium for the trivalent lanthanum in dodecahedral sites in the structure. The valency induction thus caused reduces resistivity, which is usual, but it also changes the antiferromagnetic order to a ferromagnetic order, with Curie point close to room temperature:

$$(Sr_x^{2+}La_{1-x}^{3+})_{dod}\,(Mn_{1-x}^{3+}\,Mn_x^{4+})_{oct}\,O_3^{2-}.$$

The drop in resistivity is caused by the appearance of tetravalent manganese in the octahedral sites, which both reduces local polarization by increasing the covalent character of the bond, and creates sufficient probability of transfer to the sub-level $d\gamma^1\alpha$. The change of magnetic order was attributed at the time to the appearance of another magnetic interaction mechanism (double exchange).

Next, let us consider the $H1_1$ structure, in which the superexchange mechanism predominates. In this case, the metal atoms occupying tetrahedral sites (*A*) and the metal atoms occupying the octahedral sites (*B*) often both carry magnetic moments. Two sub-structures can be distinguished:

$$\text{Normal } H1_1 \quad (T^{2+})_A(Fe_2^{3+})_B X_4^{2-}.$$
$$\text{Inverted } H1_1 \quad (Fe^{3+})_A(T^{2-}Fe^{3+})_B X_4^{2-}.$$

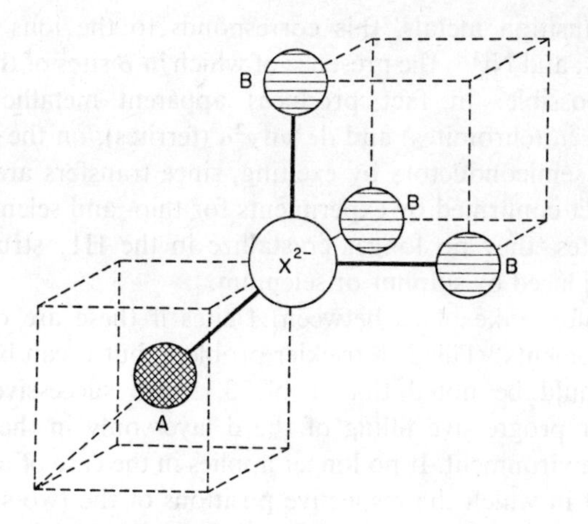

FIG. 3.11. *A*–X–*B* and *B*–X–*B* bonding angles in the $H1_1$ structure (based on Suchet, 1973).

Figure 3.11 shows that, while the short metal (*A*)–metalloid–metal (*B*) bond largely makes use of the p orbitals of the metalloid, its angle is only 125°, and that the short metal (*B*)–metalloid–metal (*B*) bond has an angle of only 90°. The superexchange interaction *B*–*B* is accordingly less than *A*–*B*, and much less than in the B1 structure. This explains the appearance of ferrimagnetism in the $H1_1$ structure, in other words the antiparallel orientation of the two ferromagnetic sub-lattices *A* and *B*. From an electrical viewpoint, in the inverted structure, the simultaneous presence of divalent iron ($d\epsilon^3\alpha d\gamma^2\alpha d\epsilon^1\beta$) and trivalent iron ($d\epsilon^3\alpha d\gamma^2\alpha$) in the *B* sites of the iron ferrite or magnetite Fe_3O_4 or of its solid solutions with other spinels naturally produces semiconduction by transfers on the sub-level $d\epsilon\beta$. The trivalent iron in the *A* sites does not give rise to any transfer.

The normal $H1_1$ structure is compatible with replacement of oxygen by sulphur and selenium. Metal–metalloid bonds thereupon become more covalent, and local polarization decreases, so that transfers in narrow bands become more frequent. Let us try to predict the compounds in which these may occur. Table 3.1 shows that transfers from site *B* to site *B* can take place for the formulae $d\epsilon^1\alpha$, $d\epsilon^2\alpha$, $d\epsilon^3\alpha d\gamma^1\alpha$, $d\epsilon^3\alpha d\gamma^2\alpha d\epsilon^1\beta$, $d\epsilon^3\alpha d\gamma^2\alpha d\epsilon^2\beta$, and $d\epsilon^3\alpha d\gamma^2\alpha d\epsilon^3\beta d\gamma^1\beta$. In the first series of transition metals, this corresponds to the ions Ti^{3+}, V^{3+}, Mn^{3+}, Co^{3+}, and Ni^{3+}, the presence of which in *B* sites of the structure, whenever possible, in fact produces apparent metallic behaviour. Formulae $d\epsilon^3\alpha$ (chromites) and $d\epsilon^3\alpha d\gamma^2\alpha$ (ferrites), on the other hand, should give semiconductors by exciting, since transfers are forbidden. This is in fact confirmed by experiments for thio- and selenochromites. As for ferrites, they no longer crystallize in the $H1_1$ structure when oxygen is replaced by sulphur or selenium.

Can transfers take place between *A* sites if these are occupied by transition elements? This is a trickier problem, but it can be answered. First, it should be noted that Table 3.1 gave successive electronic formulae for progressive filling of the d layer only in the case of an octahedral environment. It no longer applies in the case of a tetrahedral environment in which the respective positions of the two sub-levels $d\epsilon$ and $d\gamma$ are reversed. We will therefore refer to Table 3.2, which shows that, if *A–A* transfers are possible, they should be observed in the case of the ions Sc^{2+}, V^{2+}, Cr^{2+}, Fe^{2+}, Ni^{2+}, and Cu^{2+}. The formulae $d\gamma^2\alpha$ (titanium thiochromite), $d\gamma^2\alpha d\epsilon^3\alpha$ (manganese thiochromite), and $d\gamma^2\alpha d\epsilon^3\alpha d\gamma^2\beta$ (cobalt thiochromite), on the other hand, should give semiconductors by exciting, since transfers are forbidden. Experiments do not confirm these predictions. Manganese, iron, and cobalt are possible in the *A* sites of thiochromites, and all three lead to semiconductor compounds by exciting when their formula is stoichiometric. The work of Goldstein and Gibart (1969), however, indicates that the activation energy *E* of iron thiochromite is much lower, around 0.02 eV, so that a certain role may perhaps be played by the metal in site *A*.

Table 3.2. BASED ON SUCHET, 1973 (TETRAHEDRAL ENVIRONMENT)

d^1	$d\gamma^1\alpha$
d^2	~~$d\gamma^2\alpha$~~
d^3	$d\gamma^2\alpha d\epsilon^1\alpha$
d^4	$d\gamma^2\alpha d\epsilon^2\alpha$
d^5	~~$d\gamma^2\alpha d\epsilon^3\alpha$~~
d^6	$d\gamma^2\alpha d\epsilon^3\alpha d\gamma^1\beta$
d^7	~~$d\gamma^2\alpha d\epsilon^3\alpha d\gamma^2\beta$~~
d^8	$d\gamma^2\alpha d\epsilon^3\alpha d\gamma^2\beta d\epsilon^1\beta$
d^9	$d\gamma^2\alpha d\epsilon^3\alpha d\gamma^2\beta d\epsilon^2\beta$
d^{10}	~~$d\gamma^2\alpha d\epsilon^3\alpha d\gamma^2\beta d\epsilon^3\beta$~~

References

De Boer, J. H. and Verwey, E. J. W. (1937) *Proc. Phys. Soc. Lond.* **49,** 59.
Frolich (1968) *M. Dt. Akad. Wiss. Berlin* **10,** 488.
Goldstein, L. and Gibart, P. (1969) *C.R. Acad. Sci. Paris* **269 B,** 471.
Goodenough, J. B. (1963) *Magnetism and the Chemical Bond*, Wiley (Interscience), New York.
Kamigaichi, T., Hihara, T., Tazaki, H. and Hirahara, E. (1956) *J. Phys. Soc. Japan* **11,** 606.
Kjekshus, A. and Pearson, W. B. (1964) *Prog. Solid State Chem.* **1,** 83 (ed. H. Reiss), Macmillan (Pergamon), New York.
Orgel, L. P. (1960) *An Introduction to Transition Metal Chemistry*, Methuen, London, and Wiley, New York.
Suchet, J. P. (1967) *Mater. Res. Bull.* **2,** 547.
Suchet, J. P. (1969) *Annls Chim. Paris* **4,** 117.
Suchet, J. P. (1971) *Crystal Chemistry and Semiconduction in Transition Metal Binary Compounds*, Academic Press, New York.
Suchet, J. P. (1973) in *Khimicheskaya sviaz' v poluprovodnikakh i polumetallakh* (*La Liaison chimique dans les semiconducteurs et les semimétaux*) (ed. N. N. Sirota), Nauka i Tekhnika, Minsk.
Volger, J. (1961) *Semiconducting Materials* (ed. H. K. Henisch), p. 162, Butterworths, London and Washington, DC.

Chapter 4

Switching Semiconductors

4.1. Magnetic transitions

It was seen, in § 3.1, that the separation of dϵ and dγ levels into four sub-levels d$\epsilon\alpha$, d$\epsilon\beta$, d$\gamma\alpha$, and d$\gamma\beta$, in which the spin moments of electrons are directed in one direction (α) or the other (β), is very much accentuated by the existence of a ferro- or antiferromagnetic order. Because of this, it may be that these sub-levels approach bonding or antibonding level systems, so that interactions occur. In this case, one expects to observe metallic behaviour in the magnetic range, followed by semiconductor behaviour above the Curie or Néel point. This situation exists both in oxides and in more covalent compounds, and we shall give a few examples.

The manganese dioxide MnO_2 exists in several forms, of which the β phase, present in the minerals pyrolusite and polianite, is the best known. This phase has the C4 structure of rutile and a complex helicoidal antiferromagnetic order with a Néel point at 84°K. It was in 1951 that Bizette published the resistance curve in relation to temperature reproduced in Fig. 4.1. The sudden increase shown by measurements at the temperatures of liquid hydrogen, liquid nitrogen, and liquid oxygen seems to correspond to magnetic dispersion. No later measurements have been made at low temperatures, so that there is no certainty as to metallic behaviour, but this nevertheless appears possible. Above the Néel point, on the other hand, numerous instances of research have shown the semiconducting character, with $E = 0.26$ eV. This result was predictable, since the electronic formula d$\epsilon^3\alpha$ of the manganese in the octahedral site forbids any electronic transfer in the stoichiometric composition.

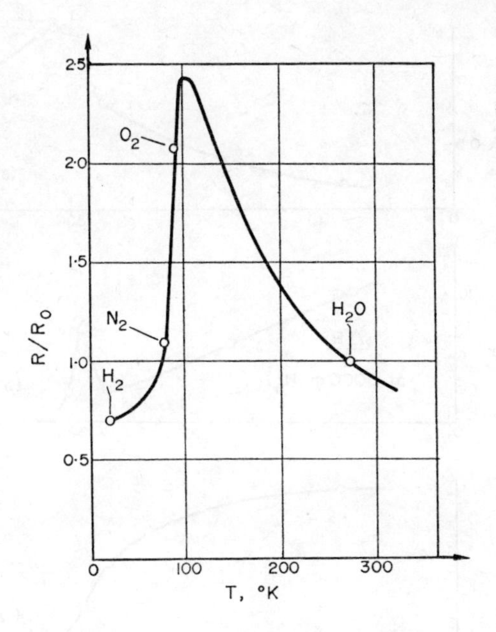

FIG. 4.1. Variation in the resistance of βMnO_2 in relation to absolute temperature (based on Bizette, 1951).

The lanthanum manganite $LaMnO_3$ has the G5 structure of perovskite. Like other manganites in this series, it is a semiconductor with high resistivity. In § 3.5 mention was made of the partial substitution of strontium and the ferromagnetism of $Sr_{0.2}La_{0.8}MnO_3$. Resistivity, considerably lowered, varies as is shown in Fig. 4.2*a*. Here, little doubt is possible: behaviour is clearly metallic at low temperatures, with strong magnetic dispersion at around 300°K, the temperature which the magnetoresistance and saturation magnetization curves in Fig. 4.2*b* and *c* identify without any ambiguity as the Curie point. At higher temperatures, subsequent research has confirmed the semiconductor character suggested by Fig. 4.2*a*. Finally, it might be mentioned that 10 years later similar behaviour was recorded in the case of substitution of 0.3 calcium atoms. In each of these two cases, the formula of the trivalent manganese is $d\epsilon^3\alpha d\gamma^1\alpha$, leaving the possibility

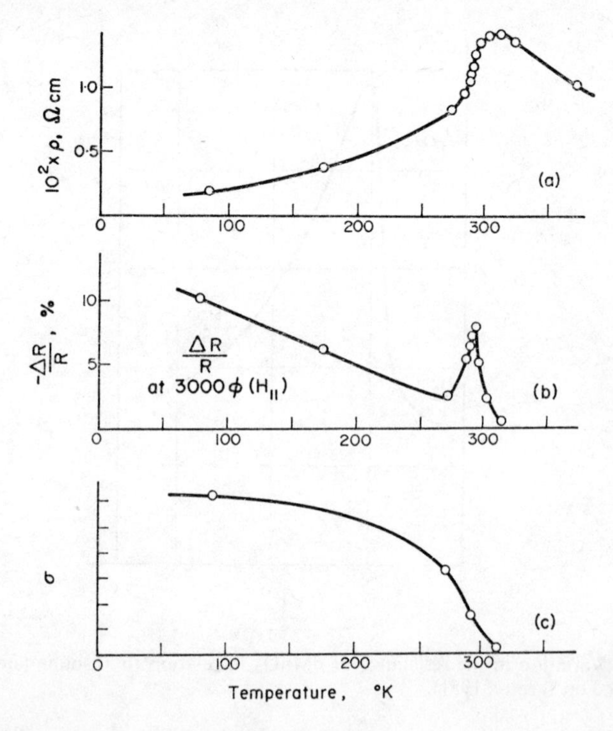

FIG. 4.2. Comparison of the variations in resistivity (*a*), magnetoresistance (*b*), and saturation magnetization (*c*) of manganite $La_{0.8}Sr_{0.2}MnO_3$ in relation to absolute temperature (based on Volger, 1954).

of transfers from Mn^{III} to Mn^{IV} or even Mn^{III} to Mn^{III}. The semiconduction mechanism therefore clearly involves transfers.

Let us now move on to compounds in which the bonds are much more covalent in character, such as antimonides with the $B8_1$ structure. The compound CrSb is antiferromagnetic, with a Néel point of 720°K. The resistivity curve in relation to temperature, published in 1957 by Suzuoka, again has the same appearance. Figure 4.3 shows a similar curve obtained by the author several years later. Magnetic dispersion is not very marked, but exists. As regards behaviour above the Néel point, it appears to be of the semiconductor type with an activation energy of 0.2 eV. The formula $d\epsilon^3\alpha$ of trivalent chromium suggests that it could involve an exciting mechanism, but as far as we know no subsequent research has yet confirmed this.

MnP compounds, with B31 structure, and MnAs and MnSb compounds with $B8_1$ structure, appear to have similar resistivity curves. They are all three ferromagnetic ($T_C = 22°$, $47°$, and $314°C$) and measurement of the ordinary Hall coefficient has shown the same transfer density for MnP and MnSb, 10^{22} per cm^3, 100 times lower than in ferromagnetic metals, and corresponding roughly to the transfer of a hole per manganese atom. These results thus correspond

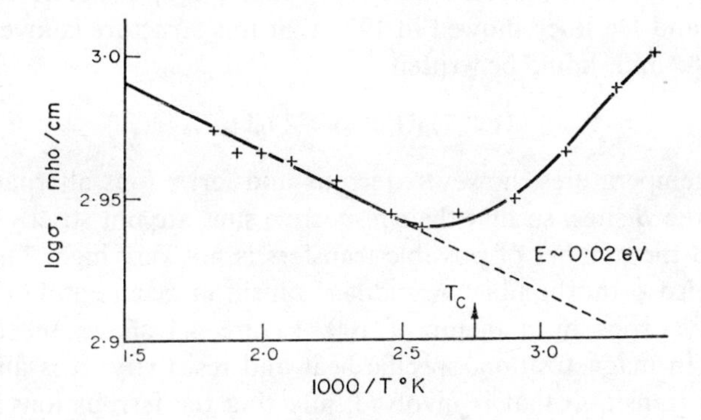

FIG. 4.3. Variation in logarithmic conductivity of CrSb in relation to the reciprocal of the absolute temperature (based on Suchet, 1963).

to what was predictable from the electronic formula $d\epsilon^3 \alpha d\gamma^1 \alpha$ of trivalent manganese and the covalent nature of the bonds: transfer occurs with zero activation energy in a $d\gamma\alpha$ narrow band. In addition, MnP is metamagnetic below 50°K for fields of more than 2 kOe, resulting in an enormous apparent magnetoresistance for all these temperatures. Finally, the resistivity curve measured by Guillaud in 1951, for MnAs, peaked clearly at the Curie point. However, it was later recognized that the problem was much more complex than it appeared, because of the temporary occurrence, between 42° and 127°C, in other words during apparent semiconductor behaviour, of a paramagnetic B31 phase.

4.2. Crystallographic transitions

This second section covers transitions involving a change in crystal structure at a temperature different from the Curie or Néel point of the principal phase. This phase change, however, may involve a magnetic transition, connected with the different arrangement of the atoms, but the transition no longer results, as in the previous section, from splitting of the d levels into α and β sub-levels.

The oldest and best-known example of crystallographic transition involving an abrupt variation in resistivity is that of magnetite Fe_3O_4, which crystallizes at room temperature in the $H1_1$ structure of spinel. Verwey and De Boer showed in 1936 that this structure is inverted for magnetite, and should be written

$$(Fe^{3+})_A(Fe^{2+}Fe^{3+})_B O_4^{2-}.$$

At low temperatures, however, ferrous and ferric ions alternate regularly in the B sites, so that their respective sites are not strictly equivalent, and the number of possible transfers is not very high. The structure is also orthorhombic. A sudden transition occurs at 120°K, and resistivity drops by a factor of 100. Figure 4.4 shows the form of changes in magnetization, specific heat and resistivity. It is an order–disorder transition that is involved, affecting the ferrous ions in the B sites of the $H1_1$ structure. Their disordered distribution above 120°K allows a much larger number of transfers, which can be evaluated by measuring the ordinary Hall coefficient at approximately 10^{20} per cm^3. Electron mobility is 0.45 cm^2/V sec, and the activation energy 0.06 eV. The transfer mechanism could have been proved, if necessary, by studying the Mössbauer effect, only an average value of the resonance of which is observed in the B sites. The exchange of electrons between ferrous and ferric ions accordingly takes place with much greater frequency than the nuclear phenomenon, which is around 100 MHz.

The previous example was not, strictly speaking, a semiconductor–metal transition, since the magnetite still has, above 120°K, semiconductor behaviour and a (transfer) activation energy of 0.06 eV despite its high conductivity. The resistivity of the oxide V_2O_3, on the other hand, a semiconductor with 0.2 eV activation energy, drops very suddenly at -100°C to 55×10^{-4} Ω cm, while all trace of activation

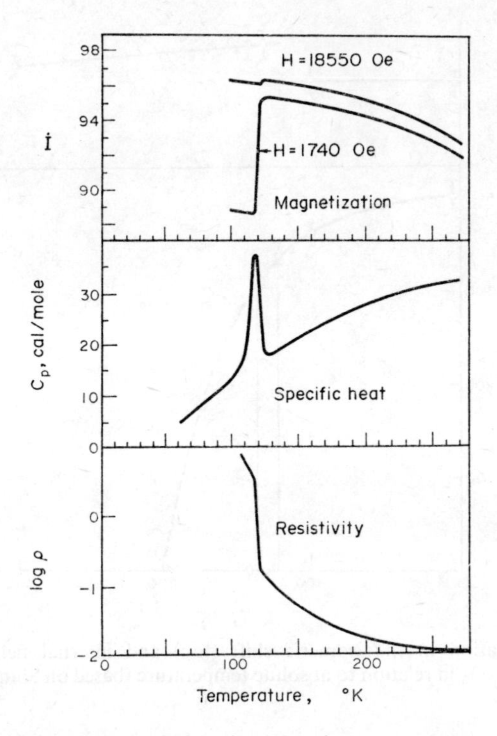

FIG. 4.4. Comparison of the variations in magnetization intensity, specific heat and logarithmic resistivity of Fe_3O_4 in relation to absolute temperature (based on Pascal, 1956–63).

energy disappears. This transition was discovered by Foëx in 1946. An antiferromagnetic order and lower crystalline symmetry exist at low temperatures, whereas above −100°C there is a change to paramagnetism and the $D5_2$ structure of corundum. The transition, however, does not occur at the Néel point. Shinjo and Kosuge succeeded in diffusing a small amount of iron enriched with isotope 57 in V_2O_3 and studying the Mössbauer effect. Figure 4.5 shows that the magnetic order disappears at about 140°K, namely −133°C, at a lower temperature than the transition, and that extrapolation of the curve of decrease of the internal field leads to a Néel point of around 200°K, namely −73°C. It is thereby established that the transition of V_2O_3 is not linked to

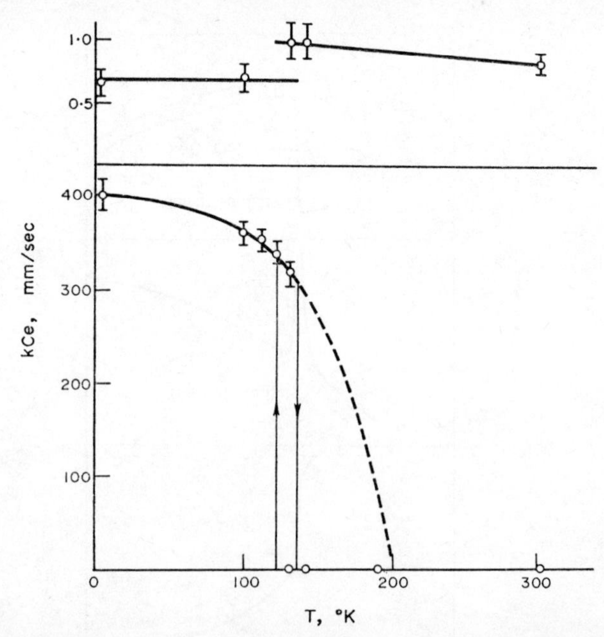

FIG. 4.5. Variation in the isomeric shift (top) and internal field (bottom) of $V_{1.98}Fe^{57}_{0.02}O_3$ in relation to absolute temperature (based on Shinjo and Kosuge, 1966).

natural disappearance of the antiferromagnetic order. At room temperature, the formula $d\epsilon^2\alpha$ is quite consistent with metallic behaviour. At low temperatures, distortion of the oxygen atom octahedron does not allow the approximation of a single $d\epsilon\alpha$ level to be maintained.

The situation is much less clear for Ti_2O_3, a semiconductor with 0.1 eV activation energy, which suddenly becomes more conducting above 200°C. The possible existence of magnetic order is a subject of much controversy here.

Finally, the oxide VO_2 crystallizes in the C4 structure of rutile, but at low temperatures shows a slight monoclinic distortion which could create alternately short and long V–V distances. In 1954 Jaffray and Dumas described its sudden drop in resistivity at 68°C. Above this temperature, its behaviour is metallic, consistent with the formula $d\epsilon^1\alpha$, and measurement of the Hall coefficient gives 3×10^{21} electrons per cm^3.

The absence of magnetic order was definitely established by diffusing iron enriched with isotope 57 and studying the Mössbauer effect. Goodenough has suggested that V–V pairs could exist in the low-temperature phase, immobilizing a d electron per V atom, and thereby in fact recreating the electronic formula $d\epsilon^0\alpha$.

In conclusion, it would appear that these so-called crystallographic transitions are fairly complex and may in fact result from direct interaction between d electrons, strong enough to modify the crystallographic structure slightly. In any case, the phenomenon appears quite marked in the case of VO_2.

4.3. Chalcogenide glasses

We shall now turn to a completely different field. Before considering the semiconductor–metal transitions of a very special type encountered in chalcogenide glasses, a few words should be said about the vitreous state in general, and about this particular category of glasses, not containing oxygen.

The arrangement of atoms in the crystal is characterized, despite inevitable defects, by the periodicity that gives rise to the crystallographic structure. The vitreous state, in contrast, involves at a long distance the same disorder as the liquid state. In fact, glass was long defined as a supercooled liquid, and is usually prepared by cooling the liquid phase. Figure 4.6 compares the volume–temperature curves obtained by cooling a liquid that can produce a crystal or glass. If cooling is slow, the basic pattern of the crystal simple, and interatomic bonds relatively ionic, crystallization usually occurs at the melting point (T_f). If, on the other hand, cooling is rapid, the basic pattern complex and bonds covalent, the slope of the curve often simply changes at T_r (*softening point*), becoming parallel to the curve of the crystal. T_r, which cannot be defined with any great accuracy, here depends on the rate of cooling.

However, it would be quite mistaken to conclude that there is a total absence of order in the atoms in the glass. The *short-distance order*, imposed by the directional character of the covalent bonds, exists as in the crystal, so that the physical properties connected with it will vary little with the change from crystal to glass if the local arrangement is

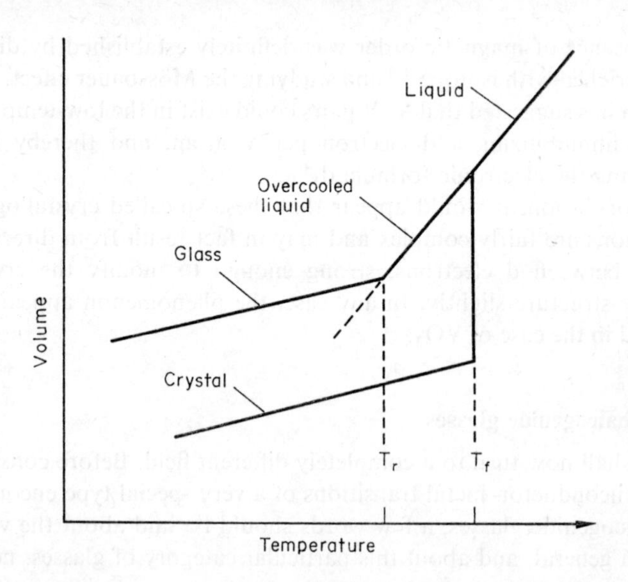

FIG. 4.6. Volume–temperature graph for a liquid liable to give both a vitreous and a crystalline solid: T_f is the melting point and T_r the softening temperature.

the same. One of these properties is semiconduction. Bloch's functions, so convenient to calculate electronic devices, in fact give a much too restrictive interpretation of the phenomenon and in no way constitute a necessary condition. Furthermore, by 1951 Ioffe had suggested that the short-distance order was solely responsible for semiconduction. This order is even more important in chalcogenide glasses, where the very covalent bonds have a very marked directional character, than in oxide glasses.

Let us start with the element selenium, in which the covalent chains —Se—Se use p atomic orbitals with an angle close to 90°, but which can vary considerably. Angular variations, added to local possibilities of rotation in relation to the bonding axis, reflect atom mobility in the liquid state. These characteristics are more or less preserved in binary glasses, such as those in the As–Se system. The system formed by the atom (or atoms connected with one another) of an element and its nearest neighbours is called a *structure unit*. For instance, a trivalent arsenic atom connects up with three selenium atoms to form a unit

$AsSe_3$, and two arsenic atoms connect up with four selenium atoms to form a unit As_2Se_4. But half the bonds of the selenium atoms are outside the unit, which should be indicated by a fractionary index in $AsSe_{3/2}$ and $As_2Se_{4/2}$. A fairly large number of structure units of crystals or glasses can be defined in this way. Glasses whose overall chemical composition corresponds exactly to that of a structure unit are thus closer to the simplicity of the crystal, and their properties are often distinguishable from those of neighbouring compositions (Suchet, 1971b).

Theoreticians have naturally tried to extend the simple models used for germanium to semiconducting glasses. For example, they have pointed out that the disorder of the atoms and the resulting potential fluctuations eliminate the sudden nature of the limit of a system of energy levels, and create a large number of localized states in the energy region which would correspond, in a crystal, to the forbidden region. Mott in Cambridge and Gubanov in Leningrad, in particular, have published a large amount of work without achieving any results of possible interest to chemists. We shall confine ourselves here to saying that it may be convenient to consider longer or shorter chains of atoms in these glasses and envisage two semiconduction mechanisms. One of these, within the chains, could show some analogy with the semiconduction mechanism by exciting germanium and the other, from chain to chain, could present some analogy with the mechanism of semiconduction by transfers in transition metal compounds. If one wants at all costs to draw a small energy diagram, partial overlapping of the bonding and antibonding level systems must be included, and the forbidden region separating them must also be filled with a large number of impurity levels.

In any case, readers will soon realize that the semiconduction mechanism in these glasses is not particularly important in order to understand the nature of the semiconductor–metal transitions they show.

4.4. Reversible crystallization

Glasses involving certain compositions crystallize more easily than others. This property is connected with the constructional simplicity of the chains, and consequently with the smallest number of structure units in the glass. Such privileged compositions generally have simple chemical formulae, like the crystals they give rise to (Suchet, 1971b). If the conductivity of the crystal is much higher than that of the glass, the passage from glass to crystal will be revealed by the appearance of a highly conducting state, and return to the vitreous state by the appearance of a relatively insulating state. The high conductivity of the crystal, in fact, simply results from its large impurity content, since the initial glass does not generally have either the purity or the exact composition that would allow an intrinsic or even fairly pure semiconductor crystal to form. This feature can, what is more, be deliberately accentuated by introducing a slight variation in composition or a foreign element.

The first publications containing a clear description of the phenomenon with its two stable states, and mentioning its usefulness in the construction of memories, appear to be those of Pearson and his collaborators in 1962 (Pearson, 1964), and Kolomiets and Lebedev in 1963, although certain patent applications date from before. The low thermal conductivity of chalcogenide glasses in which it occurred soon led to the hypothesis of a thermal phenomenon connected with a very localized phase change. Fritzsche, studying a $Ge_{16}Te_{82}Sb_2$ glass, close to the eutectic $Ge_{15}Te_{85}$, in 1969, carried out a differential thermal analysis, and obtained the results shown in Fig. 4.7. Curve *a* represents heating: it shows the softening point T_r, crystallization point T_1, and melting point T_2. Curve *b* represents fast cooling, restoring the glass. Curve *c* represents slow cooling with solidification at temperature T_3, close to T_2, and the obtaining of crystals. Curve *d* represents the heating of the crystals obtained during *c*, with the melting point T_2 reoccurring. This experiment, involving a composition very close to a crystallizable eutectic with a simple formula, leaves very little room for doubt.

Almost simultaneously, another research worker applied a voltage of 150 V to the surface of a $Ge_{12}As_{19}Te_{69}$ glass, between two electrodes 3 mm apart. Where the passage of the current was sufficiently short in duration, he obtained a fine filament that melted and quickly solidified in crystal form. What happens on a surface of a glass can also take place inside a thin strip, and the crystallization process accordingly seems clearly established, at any rate for glass compositions that are easy to crystallize. Which are these? Two examples have already been seen:

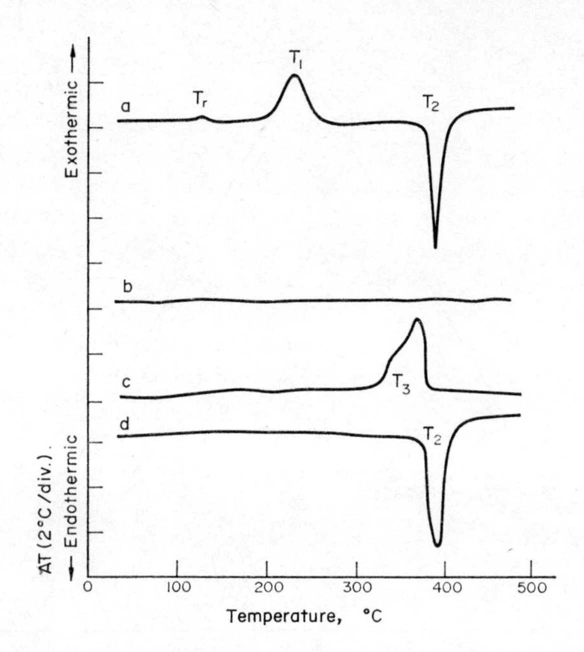

FIG. 4.7. Curves of thermal differential analysis of the composition $Ge_{16}Te_{82}Sb_2$ (based on Fritzsche, 1969). Reprinted with permission of the publisher; copyright 1969 by International Business Machines Corporation.

one is the composition of a crystallizable binary, to which has been added 2% atoms of an impurity that can strongly increase the electronic conduction of the crystal formed; the other is a ternary composition situated on the edge of a vitreous region of the phase diagram.

The origin of this crystallization will be better understood from consideration of Fig. 4.8, which indicates the temperatures of the start of crystallization and crystallization isotherms in the Ge–As–Se ternary system. There are five crystallization poles in the vitreous region of this diagram: the element selenium, binary compounds $GeSe_2$, As_2Se_3, and AsSe, and ternary compound GeAsSe. Complete, homogeneous crystallization therefore appears obtainable only with simple compositions, comprising a single structure unit, and the chemical formula of which satisfies Dalton's laws on defined compounds. For neighbouring compositions, a mixture of two crystallized phases is obtained, or, more

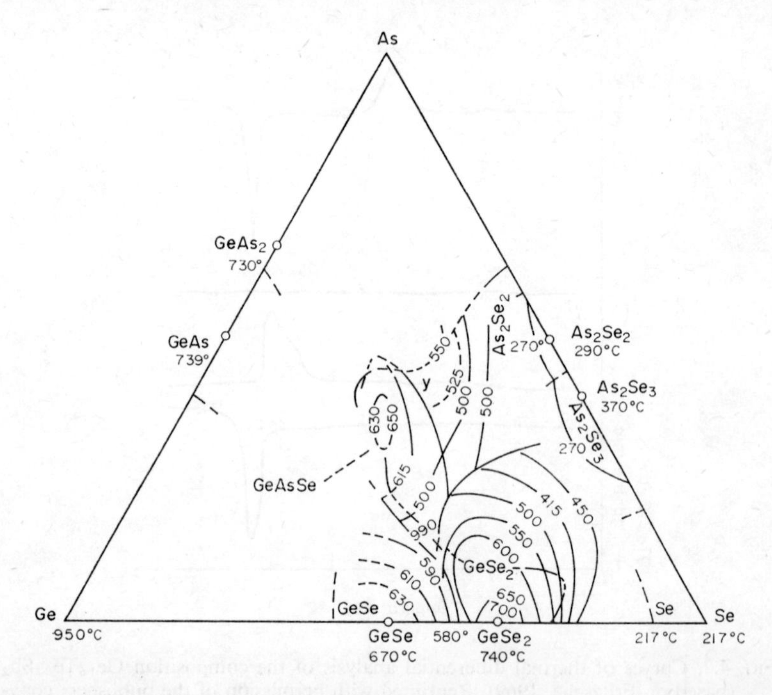

FIG. 4.8. Temperatures of start of crystallization and crystallization isotherms in the Ge–As–Se system (based on Vinogradova *et al.*, 1968).

frequently, of a crystallized phase and vitreous phase. One method of finding these crystallizable glasses therefore consists of remaining close to a natural crystallization pole.

The second method concerns compositions situated on the limits of vitreous regions, and crystallization of which can always occur if cooling is too slow. It raises the problem of predicting such limits in diagrams that are still not well known. Such prediction is very tricky, since glass formation mechanisms are still a subject for argument. In the case of ternary diagrams based on two known binary glasses, however, it is possible to show that the limits of these regions are very approximately defined by the rules of substitution of the third element in the structure unit or units of each binary glass, and by the rules of exclusion of structure units foreign to the two binary glasses. In the Ge–As–Se ternary diagram one thus obtains two separate zones, one corresponding to the substitution of arsenic for selenium in the $GeSe_{4/2}$ unit, and the other corresponding to the substitution of germanium for selenium in the $AsSe_{3/2}$ unit (Suchet, 1971b).

4.5. Non-destructive breakdown

Right at the beginning of this book, in § 1.2, reference was made to what happens when the voltage applied to two electrodes enclosing a solid insulating sample is gradually increased. It was pointed out that, if the increase in voltage is faster than the development of secondary phenomena such as temperature rise and melting, breakdown is inevitable when the field reaches approximately 1 MV/cm.None of the theories so far produced have succeeded in explaining simultaneously the breakdown that occurs in all media in which it has been observed, so that different physical phenomena appear capable of causing it. In crystallized solids, it is nearly always destructive, with melting or decomposition of the material along the path followed by the discharge. In liquids, on the other hand, the mobility of the atoms allows the material to reform after the discharge. Oxide glasses, often used for their dielectric properties, are subject to destructive breakdown, like crystals, but the particular behaviour of agglomerates of grains of silicon carbide (non-linear resistors) has sometimes been attributed to breakdown of the very fine surface layer of silica which covers them, and in this case there would be non-destructive breakdown. As for chalgogenide glasses, 10 years ago they already appeared to show different behaviour, suitable for electronic applications.

Before going any further, it must be pointed out that sudden variations in conductibility, with the appearance of a short circuit, may be due to widely varying phenomena. Two metal electrodes, placed close together and subjected to an intense electric field, may present accidental contact, the substance separating the electrodes may be decomposed irreversibly, and reactions may take place with the surrounding medium or electrode metal. Such phenomena are extremely common but do not produce reversible, reproducible effects. The same early publications mention such effects and reversible crystallization in chalcogenide glasses, since both causes, although different, give similar effects, and were confused in the beginning. The description given notably by Pearson and his collaborators of their observations on As–Te–I glasses is correct, and corresponds to the effect analysed below, but this material did not allow satisfactory reproducibility, so that their research could not make any progress (Suchet and Maghrabi, 1972).

Ovshinsky, after studying such non-linear effects in various materials, turned to chalcogenide glasses at the beginning of the sixties, and performed experiments on a very large number of different compositions. It was only in 1968, after testing the reproducibility of the effect shown by the glass $Si_{12}Ge_{10}As_{30}Te_{48}$, that he published a description. A thin film, 0.5 μm thick, between carbon electrodes, shows a resistivity at room temperature of 20 MΩ cm for low electric fields, but suddenly, in less than 0.15 nsec (nanoseconds), takes on similar conductivity to that of metals if subjected to voltage higher than a threshold-voltage of approximately 14 V. Oscillograms of responses to a sinusoidal current with a frequency of 60 Hz and to square impulses establish its total reversibility. This effect is illustrated in two figures. Figure 4.9

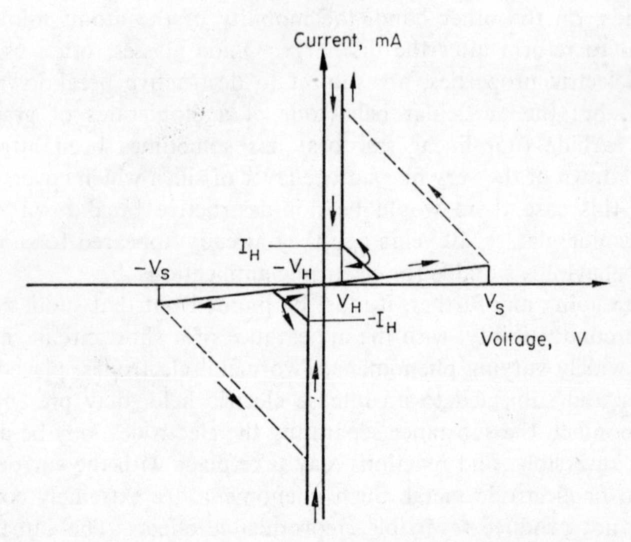

FIG. 4.9. Ideal I/V characteristics for a chalcogenide glass under voltages close to its threshold (based on Csillag *et al.*, quoted by Suchet and Maghrabi, 1972).

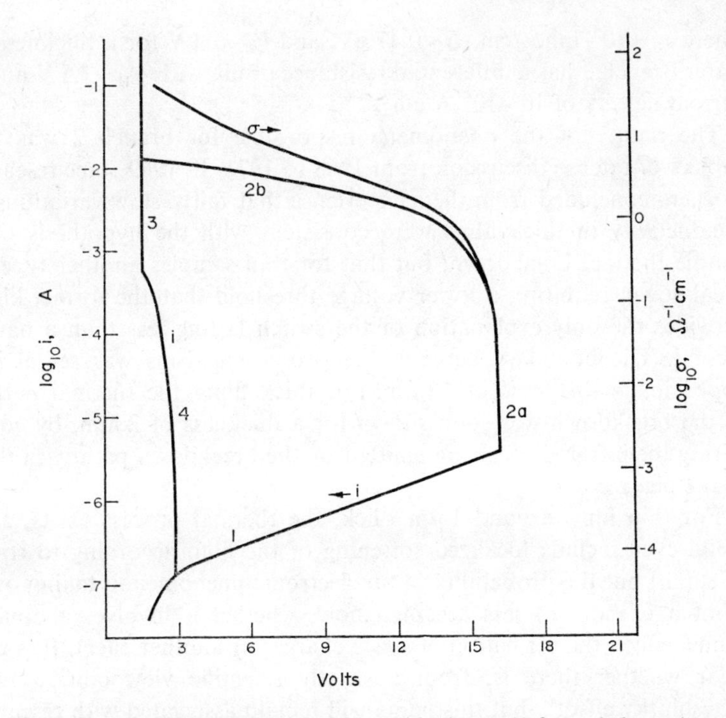

FIG. 4.10. Real I/V characteristics (semi-logarithmic graph) for a chalcogenide glass under voltages close to its threshold (based on Böer *et al.*, quoted by Suchet and Maghrabi, 1972).

represents a characteristic ideal I/V curve, which defines a number of simply measured parameters: voltage threshold V_S, minimum voltage V_H and minimum current I_H. Figure 4.10 is a semi-logarithmic representation of a real characteristic I/V curve for the glass $Ge_{12}As_{30}S_{25}Se_1Te_{22}V_{10}$: branch 1 insulating, 2*a* temperature rise, 2*b* breakdown, 3 conducting, 4 return to the initial state. The conductivity σ is given in the same coordinates for branch 2.

In both figures, certain parts of the curve are obviously determined by the circuit in which the glass is inserted. In the second figure, Böer *et al.* indicate that conductivity in branch 1 follows the law

$$\sigma \sim \sigma_0 \exp[-(E/kT) + (V/V_0)],$$

where $\sigma_0 \sim 10^3$ mhos/cm, $E \sim 0.47$ eV, and $V_0 = 4$ V for a thickness of 1 μm. Branch 3 has a differential resistance of nil, with $V_H \sim 1.5$ V and a current density of 10^5–10^6 A/cm^2.

The nature of the phenomena responsible for branch 2 was the subject of intense discussion from 1968 to 1971. In 1969 some research workers concluded from their experience that fairly slow variations in conductivity in thick films were consistent with the hypothesis of a simple thermal breakdown, but that for thin samples another type of breakdown, requiring a lower voltage threshold than the former kind, provides the only explanation of the switch lasting less than a nanosecond. The boundary between the two mechanisms was set at the approximate thickness of 10 μm. For thick films, the thermal nature of the breakdown was soon shown for a thickness of 2 mm, by comparing the infrared spectrum emitted by the breakdown path with that of the black body.

For thin films around 1 μm thick, the thermal process exists, and could even include localized softening of the film, according to some research; but the probability of an electronic phenomenon taking over from it is more or less accepted now, whether it involves a double tunnel effect, the formation of space charges or another effect. It is not clear whether there is, from a strictly scientific viewpoint, a new "Ovshinsky effect", but this name will remain associated with research into materials presenting a reproducible effect.

References

BIZETTE, H. (1951) *J. Phys. Radium* **12,** 161.
FOËX, M. (1946) *C.R. Acad. Sci. Paris* **223,** 1126.
FRITZSCHE, H. (1969) *IBM J. Res. Devt.* **13,** 515.
GOODENOUGH, J. B. (1960) *Phys. Rev.* **117,** 1442.
GUILLAUD, C. (1951) *J. Phys. Radium* **12,** 223.
IOFFE, A. F. (1951) *Izv. Akad. Nauk SSSR, Ser. Fiz.* **15,** 477.
JAFFRAY, J. and DUMAS, D. (1954) *J. Rech. CNRS Bellevue, Paris* **27,** 360.
KOLOMIETS, B. T. and LEBEDEV, E. A. (1963) *Radio Eng. Elektron. USSR* **8,** 1941 and 2097.
OVSHINSKY, S. R. (1968) *Phys. Rev. Lett.* **21,** 1450.
PASCAL, P. (1956–63) *Nouveau traité de chimie minérale*, Masson, Paris.
PEARSON, A. D. (1964) *Modern Aspects of the Vitreous State* **3,** 29 (ed. J. D. Mackenzie), Butterworths, London.
SHINJO, T. and KOSUGE, K. (1966) *J. Phys. Soc. Japan* **21,** 2622.
SUCHET, J. P. (1963) *Annls Phys. Paris* **8,** 285.

SUCHET, J. P. (1971a) *Crystal Chemistry and Semiconduction in Transition Metal Binary Compounds*, Academic Press, New York and London.
SUCHET, J. P. (1971b) *J. Non-Cryst. Solids* **6,** 370.
SUCHET, J. P. and MAGHRABI, C. E. (1972) *Annls Chim. Paris* **7,** 157.
SUZUOKA, T. (1957) *J. Phys. Soc. Japan* **12,** 1344.
VERWEY, E. J. W. and DE BOER, F. (1936) *Recl. Trav. chim. Pays-Bas Belg.* **55,** 531.
VINOGRADOVA, G. Z., DEMBOWSKII, S. A. and LUZHNAYA, N. P. (1968) *Zh. neorg. Khim.* **13,** 1444.
VOLGER, J. (1954) *Physica* **20,** 49.

Chapter 5

Insulators

5.1. Inorganic insulators

We saw, in § 1.2, that in a system of atoms connected with one another the only possible insulators at average temperatures resulted either from strongly ionic interatomic bonds or strongly covalent interatomic bonds with, in the latter case, relatively light atoms. In this section we shall consider ionic (inorganic) insulators and in the next section organic insulators, which constitute the main body of the second category. It should also be pointed out that molecular solids, in which a very large number of systems of atoms exist, more or less independent of one another, are particularly good insulators. They are to be found almost only among organic compounds.

The disadvantage of inorganic insulators is that they involve rather difficult working and machining because of their hardness, which is much greater than that of metallic conductors. Consequently, they replace organic insulators only if they offer other vital properties such as a higher breakdown voltage (disruptive strength) or greater temperature resistance (refractoriness). The most widely used insulators are oxides of light elements in columns II–IV of the Periodic Table. Alkalines on the one hand, and heavy metals on the other, have ions of very different dimensions from those of oxygen, and result in types of atomic arrangement conducive to ionic semiconduction. Since glucine is ruled out because of its cost and toxicity, there remain MgO, B_2O_3, Al_2O_3, SiO_2, and GeO_2. Certain very refractory transition metal and lanthanide oxides can also occasionally be used.

Massive crystals are seldom employed because they are difficult to handle. The quartz SiO_2 is sometimes used, despite its high cost, but

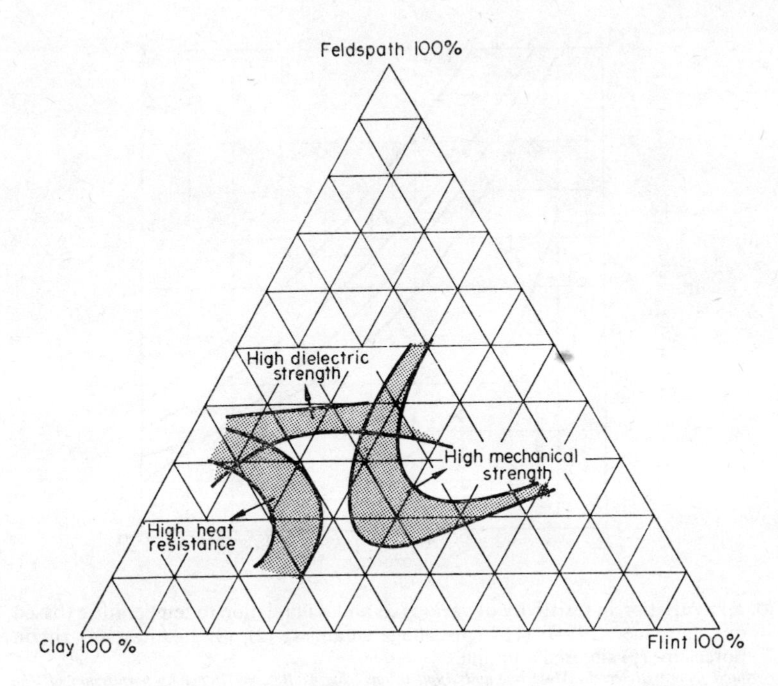

FIG. 5.1. Compositions of the porcelains with various properties in the diagram clay–feldspath–silica (based on Salmang, 1951).

the most successful have been micas, either in their natural form, in which they split easily into thin sheets, or in the form of powder, agglomerated by an asphalt-based binder (micanites). The disruptive strength of the sheets decreases in relation to thickness. In mycalex, used in high-frequency equipment since the Second World War, the binder consists of lead borate brought to the vitreous state at around 750°C, and moulding is done at high temperature under pressure. Other compounds are generally used in porcelains (MgO, Al_2O_3, SiO_2) or glasses (B_2O_3, SiO_2, GeO_2), which can be formed by ordinary ceramic methods or melting. Raw materials employed always involve the three substances in the diagram in Fig. 5.1: clay or hydrated aluminium silicate $Si_2O_3Al_2(OH)_4$, which confers plasticity, feldspath, or aluminium alkaline silicate $Si_3AlO_8Na(K)$, which acts as binder, and quartz SiO_2. After firing, the final product contains crystals of quartz SiO_2

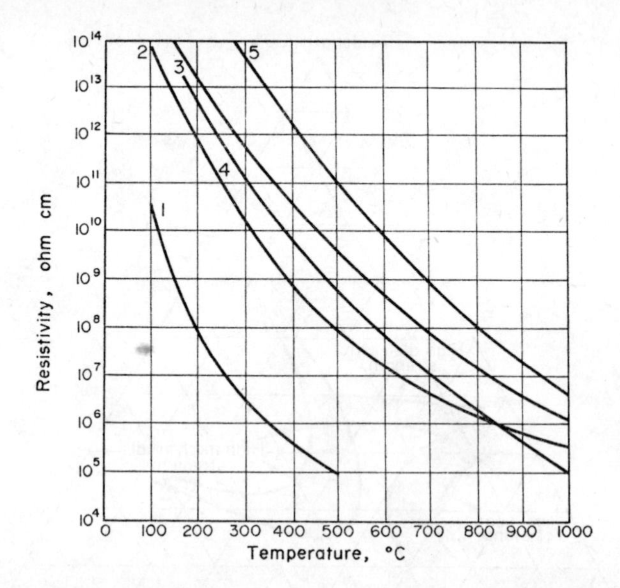

FIG. 5.2. Variation in resistivity of various ceramics in relation to temperature (based on Thurnauer, 1954): (1) high-voltage ceramics; (2), (3) steatites; (4) zircon porcelain; (5) sintered alumina.

Reprinted from "Dielectric Materials and Applications" by A. R. von Hippel by permission of The M.I.T. Press, Cambridge, Massachusetts.

and mullite $Si_2Al_6O_{13}$ coated with a vitrified material coming from feldspath. Other porcelains contain magnesia: steatite SiO_3Mg, produced from talc and clay, refractory forsterite SiO_4Mg_2 (1900°C), cordierite $Si_5O_{15}Mg_2Al_2$ with a little aluminium, etc. For high-voltage insulators, moisture absorption must be nil. Figure 5.2 shows the variations in the resistivity of various ceramics in relation to temperature, and Table 5.1 shows their main properties.

At high temperatures, resistance to thermal impacts and refractoriness play a more important role. The first of these qualities leads to the elimination of porcelains containing a vitreous binder, which would be subject to deformations, and the adoption of porcelains containing alumina, zircon, or lithia. The second quality is more complex, and, apart from ceramics with a high alumina content, also used to construct kilns and melting pots, one may need to turn to specifically refractory products, notably oxides of metals containing empty or completely full d or f sub-levels, whether sub-levels $d\epsilon\alpha$ and $d\gamma\alpha$ (MgO, 2800°C

Table 5.1. BASED ON THURNAUER, 1954

	Maximum temperature of use (°C)	Disruptive strength (kV/mm[a])	Room temperature resistivity (Ω cm)
High-voltage porcelain	1000	100–160	10^{12}–10^{14}
Alumina porcelain	1350–1500	100–160	10^{14}–10^{15}
Steatite	1000–1100	80–140	10^{13}–10^{15}
Forsterite	1000–1100	80–120	10^{13}–10^{15}
Zircon porcelain	1000–1200	100–140	10^{13}–10^{15}
Lithia porcelain	1000	80–120	
Low-voltage porcelain	900	16–40	10^{12}–10^{14}
Cordierite refractories	1250	16–40	10^{12}–10^{14}
Aluminium silicate refractories	1300–1700	16–40	10^{12}–10^{14}

[a] Thickness 6.35 mm.

Reprinted from Dielectric Materials and Applications by A. R. von Hippel by permission of The M.I.T. Press, Cambridge, Massachusetts.

and CaO, 2600°C, $d\epsilon^0\alpha$; Cr_2O_3, 2400°C, $d\epsilon^3\alpha$; NiO, 2500°C, $d\gamma^3\alpha$) or $f\alpha$ sub-levels (ThO_2, 3000°C, $f^0\alpha$; Gd_2O_3, 2400°C, $f^7\alpha$). In 1969 we showed that these oxides offered maximum refractoriness.

Finally, glasses are good insulators in the mass, but care must always be taken with regards to their surface conduction in damp atmospheres, notably for alkaline glasses. For instance, ordinary sodo-calcic glasses have a surface resistance, expressed in ohms per square, of 10^{13} in dry air, but of only 10^9 above 50% humidity. The position is better for Pyrex glasses, which are more refractory. The conditions of their use, in competition with ceramics, often lead to the choice of glasses with good mechanical strength, or even hardened glasses. In general, it may be noted that electrical conduction in glasses is, in most cases, ionic, and is thus connected with refractoriness. However, there are certain compositions and temperature ranges for which electronic conduction has been observed in the mass of the glass. These are always special glasses, containing transition metals that can give ions of multiple valencies. They are generally based on $V_2O_5+P_2O_5$, $Fe_3O_4+P_2O_5$, or $Fe_2O_3+B_2O_3$.

Earlier chapters dealt with the conditions in which greater electrical conduction could appear in inorganic solids, and described three semiconduction mechanisms: exciting (germanium), transfer on a d level (mixed-valency, crystallized, or vitreous compounds), and the still unfamiliar mechanism of chalcogenide glasses.

5.2. Organic insulators

The formula of all organic insulators contains carbon atoms, and they always carbonize at a fairly low temperature. Use of them has spread considerably in electrical engineering because they are low in cost and easy to mould and machine. All of them result from condensation reactions with the formation of giant molecules (polymerization). They are referred to under the general name of "plastics". Their composition is usually at an intermediate stage between two ideal types: linear or filiform molecule and three-dimensional or latticed molecule.

Many *thermoplastic* materials resemble the first of these two types. In such substances, several thousand atoms are linked together in a single direction by strong covalent bonds. For example, a molecule of polyisobutylene, which has a molecular weight of 250,000, consists of a chain of 8000 carbon atoms, but its average thickness corresponds to only two of these atoms. Possible deformations of such a chain obviously play a large part in the mechanical properties of such substances, the plasticity of which increases with rising temperature (hence their name). *Thermosetting* materials, as well as processed rubbers, on the other hand, resemble the second type, in which there are bonds between the chains, forming a three-dimensional mesh. If the mesh is slack, the mechanical properties remain close to those of the first type (natural rubber). If, on the other hand, it is tight, the substances become rigid, and sometimes even brittle. The effect of heat initiates or completes their polymerization, causing them to set harder (hence their name). For example, products of condensation between phenols and aldehydes, such as phenolformaldehydes, are good electrical insulators, used either in a thin layer, or with inert fillers to obtain moulded components. Figure 5.3 shows the formation of phenolformaldehyde polymer in diagrammatic form.

The simplest linear polymers are polythene $(—CH_2—)_n$, polytetrafluoroethylene $(—CF_2—)_n$ or "Teflon", polyvinyl chloride $(—CH_2—CHCl—)_n$ or PVC, and polystyrene $(—CH_2—CHC_2H_5—)_n$. More

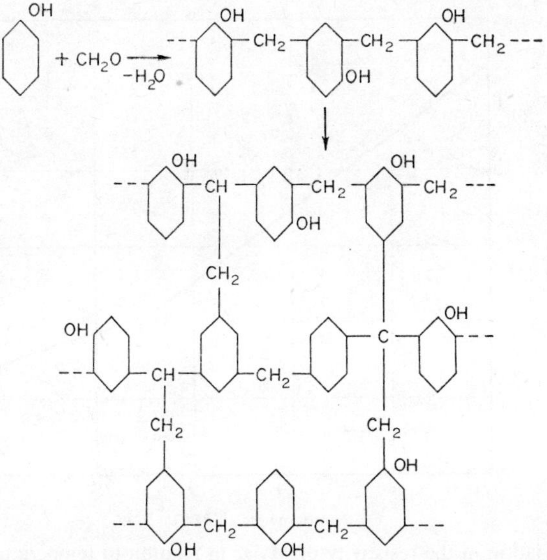

FIG. 5.3. Formation of the polymer of phenolformaldehyde (based on Busse, 1954). *Reprinted from "Dielectric Materials and Applications" by A. R. von Hippel by permission of The M.I.T. Press, Cambridge, Massachusetts.*

complex polymers include polyesters, resulting from the condensation of diacids and glycols, e.g. terephthalic acid and ethylene-glycol (Dacron in the United States, Terylene in Britain, Tergal in France, Mylar in thin layers), polyamides, resulting from the condensation of diacids and diamines (nylon family), natural rubber, silicone resins $(\text{—O—SiR}_2\text{—})_n$, where R is any organic radical, etc. Each of them has its own qualities: elasticity for rubber (but low ozone resistance), mechanical properties for PVC and nylon (but poor electrical resistance), fairly good thermal stability (250°C) for Teflon, insulating qualities for polythene protected from sunlight (but softening at 85°C), and Mylar, etc. Figure 5.4 shows the variation in resistivity in relation to temperature for Mylar: its softening point is 240°C, its disruptive strength is 20 kV/mm, and it has a resistivity of 4×10^{15} Ω cm at room temperature. If these figures are compared with those on Table 5.1, they seem poor at first sight, but Mylar can be produced in films only a few microns thick.

Among three-dimensional polymers, mention has already been made of phenolformaldehyde. There is also a polyethylene derivative:

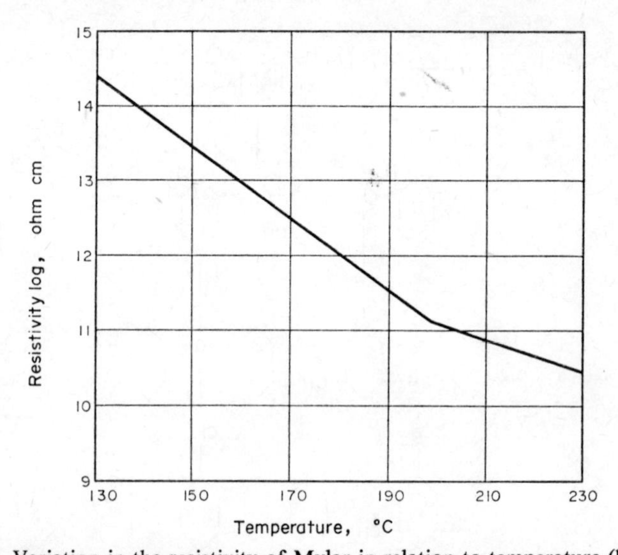

FIG. 5.4. Variation in the resistivity of Mylar in relation to temperature (based on Busse, 1954).
Reprinted from "Dielectric Materials and Applications" by A. R. von Hippel by permission of The M.I.T. Press, Cambridge, Massachusetts.

chlorosulphonated polyethylene or Hypalon, which contains one SO_2Cl group per 90–100 carbon atoms. The addition of small quantities of oxides such as MgO or PbO catalyses the formation of a three-dimensional lattice that is remarkably resistant to oxidation by ozone, which is important at high voltages, where effluvia cause formation of this gas (resistance of more than 100 hours, compared with 3 minutes for natural rubber in an atmosphere containing 130/1000 ozone).

In the case of organic insulators, the appearance of higher electrical conduction in certain categories of substances is governed by much more complex factors than for inorganic insulators. One can begin by distinguishing three categories of "organic semiconductors" in which conductivity is at least approximately the same as for distilled water. These are pyropolymers, in which the structure imperfectly reproduces that of graphite (up to 100 mhos/cm), charge transfer complexes, in which aromatic carbides and inorganic ions are associated (10^{-2}–10^{-9} mhos/cm), and stable anion-radicals, which can be isolated in the crystallized state (approximately 1 mho/cm).

It has long been known that the mechanism of pyrolysis increases the number of cycles and double bonds in polymers. This results from research into electron paramagnetic resonance and ultraviolet absorption. Toptchiev *et al.* (1959) showed, for example, that heat treatment of polyacrylonitrile creates double bonds conjugated by oxidation (Fig. 5.5). Under certain conditions, a semiconductor is obtained that is stable up to 300°C and has an activation energy of 1.7 eV. The conduction mechanism appears to be connected with the presence of π electrons in the cycles. In the second category, inorganic ions, usually alkalines or halogens, lose or gain an electron from the aromatic carbide, to give an association of the type $[R]^{-}Na^{+}$ or $[R]^{+}Cl^{-}$ (Suchet,

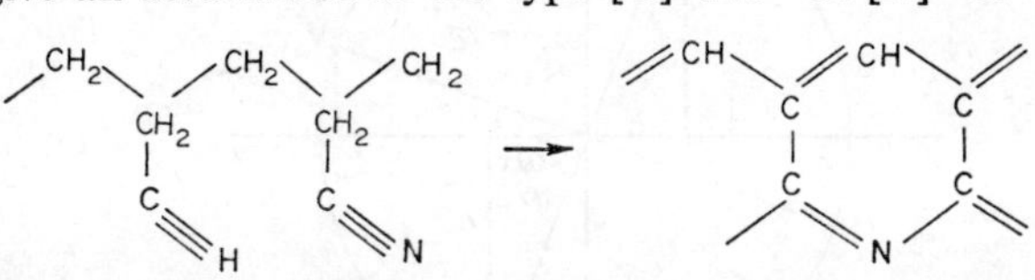

FIG. 5.5. Appearance of conjugated double bonds in polyacrylonitrile (based on Toptchiev *et al.*, 1959).

1962). The pre-exponential term σ_0 of the usual relation of semiconductors, probably connected with the passage of carriers from one molecule to the other, is then greatly reduced, and conductivities of around 1 mho/cm are even reached with an activation energy of several hundredths of an electron volt for the complex perylene-iodine. The third category has recently aroused most interest. Tetracyanoquinodimethane (TCNQ), in particular, is a quinonic reagent made strongly electrophile by two dicyanomethylene substituents situated in the para position (Néel and Dupuis, 1972). In capturing an electron, it gives rise to the anion-radical $TCNQ^{-}$, in which two charges, approximately equal to 0.5 electronic unit, are separated by a distance of about 5.4 Å. This results in an external field which allows the ion-radical to associate very stably with a neutral molecule, namely ($TCNQ$, $TCNQ^{-}$).

5.3. Alternating currents

So far, we have mainly considered electrical conduction caused by the application of a d.c. potential difference between two metal elec-

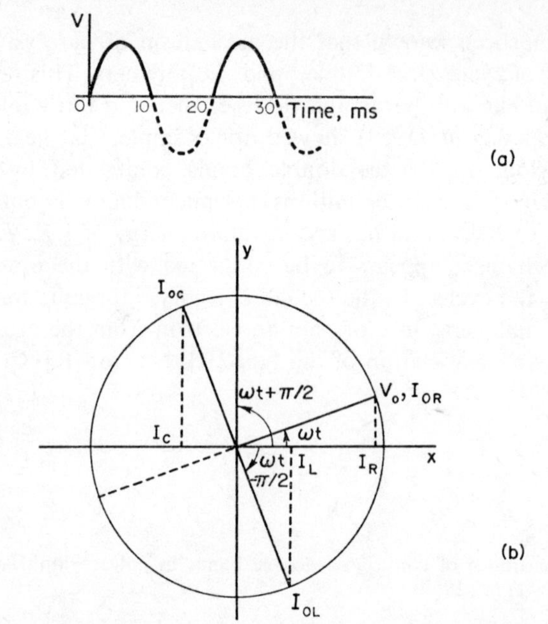

FIG. 5.6. Variation in an a.c. potential in relation to time (*a*) and its representation by a rotating vector (*b*).

trodes. The behaviour of solid materials is not the same in the presence of an a.c. potential difference, and materials which are insulating in the former case can under some conditions become conducting. Let us assume that we have a source of electricity such that the potential applied to the electrodes, in relation to time, is as shown in Fig. 5.6. Voltage is nil at a certain instant 0, passes through a maximum after 5 msec, is nil again after 10 msec and becomes inverted, is minimal after 15 msec, again is nil after 20 msec (standardized industrial frequency $\nu = 50$ periods/sec or 50 Hz). It can be represented by the projection on the abscissae axis of a rotating vector with same origin. If ω is the angular velocity $2\pi\nu$ (100π in the case above) of the vector and t is the time, the algebraic value of this voltage can be written $V = V_0 \sin \omega t$ and the current intensity it produces in a resistance R is similarly $I = I_0 \sin \omega t$ with $I_0 = V_0/R$: the two quantities V and I are in phase, in other words they increase and decrease simultaneously.

Depending on the nature of the materials subjected to a.c. potential and the form of the circuit in which the charges are moving, however, the state of affairs can change. One then refers to a capacitive effect if the current I is in advance of, and an inductive effect if it is behind, the voltage V. The concept of electrical resistance is generalized by that of *impedance* Z, in which one distinguishes resistance $Z = R\,(\mathbf{V}, \mathbf{I} = 0)$, capacitance $Z = 1/C_0\omega\,(\mathbf{V}, \mathbf{I} = \pi/2)$ and inductance $Z = L_0\omega\,(\mathbf{V}, \mathbf{I} = -\pi/2)$. These last two quantities are functions of ω, namely of the frequency used. The current produced by the same voltage V_0 in a purely capacitive or purely inductive circuit could be written $I_C = I_{0C} \sin(\omega t + \pi/2)$, namely $-I_{0C} \cos \omega t$, with $I_{0C} = V_0 C\omega$, or $I_L = I_{0L} \sin(\omega t - \pi/2)$, namely $I_{0L} \cos \omega t$, with $I_{0L} = V_0/L\omega$. If the circuit simultaneously comprises a capacitance and an inductance in series, the capacitance has to be given the minus sign as follows: $I_{CL} = I_{0\,CL} \cos \omega t$, with $I_{0\,CL} = -V_{0C}C\omega = V_{0L}/L\omega$. Since $V_0 = V_{0C} + V_{0L}$, then

$$I_{0\,CL} = V_0/(L\omega - 1/C\omega).$$

The quantity $(L\omega - 1/C\omega)$ is called *reactance* and is represented by the symbol X.

It might be pointed out that in a particular case in which $LC\omega^2 = 1$, one obtains $I_{0\,CL} = +\infty$, while infinite overvoltages appear at the terminals of the capacitance ($V_{0C} = +\infty$) and inductance ($V_{0L} = -\infty$). This remarkable phenomenon is the *resonance* of the circuit for the particular frequency $\nu = \frac{1}{2}\pi\sqrt{(LC)}$ thus defined. In practice, the circuit always contains a certain resistance above zero, which has to be taken into account and which restricts the values reached by I_{CL}, V_C, and V_L. One then has

$$I_{0\,CLR} = V_0/Z = V_0/\sqrt{[R^2 + (L\omega - 1/C\omega)^2]}$$

and the value $LC\omega^2 = 1$ is reflected only by maximum current I_{CLR} in relation to neighbouring frequencies, increasing in acuteness in inverse ratio to R. The same applies to overvoltages. This is the resonating circuit. If capacitance and inductance are in parallel, it is easily shown that there is a contrasting minimum current I_{CLR} in relation to neighbouring frequencies. This is the loop circuit.

Let us return to representation of the rotating vector in the practical case of a real capacitance, in other words comprising a parallel resistance resulting from the fact that the medium constituting it is not a perfect insulator. The current passing through it results from the addition of $I_{0C} = V_0 C\omega$ with angle $\omega t + \pi/2$ and of $I_{0R} = V_0/R$ with angle ωt. Its phase angle can therefore be evaluated by $\omega t + \varphi$, so that $0 < \varphi < \pi/2$, with

$$\tan \varphi = RC\omega.$$

This tangent, generally expressed as R/X (parallel resistance) or X'/R' (series resistance), is also called the quality or overvoltage factor Q of the condenser formed by both electrodes and the medium separating them. The *loss angle*

$$\delta = \pi/2 - \varphi$$

is also defined; its tangent, consequently the reciprocal of the preceding one, is also called the dissipation factor $1/Q$. Here it equals $1/RC\omega$ (or X/R for a parallel resistance) and plays a large part in assessing losses at high frequencies.

Let us now look a little more closely at what happens in a capacitance. The typical circuit element is the plane condenser represented in Fig. 5.7. The *electric field E* is normally defined as the average field in a long, needle-shaped cavity along the direction of polarization (generally perpendicular to the two plane electrodes). The *electric displacement D* is similarly defined as the average field in a disc-shaped cavity the plane of which is perpendicular to the direction of polarization (generally parallel to the two plane electrodes). The difference

$$D - E = 4\pi P$$

results from the field $4\pi P$ of the density P of charges appearing on the plane surfaces of the disc-shaped cavity (the density corresponding to the needle-shaped cavity can be ignored). We shall return in the next section to the practical significance of P. In the figure, the voltage applied to the condenser is E times its thickness w, and the thin layers of air that may separate the condenser material from its two plane electrodes are ignored.

The ratio D/E is the specific inductive capacity or *dielectric constant*

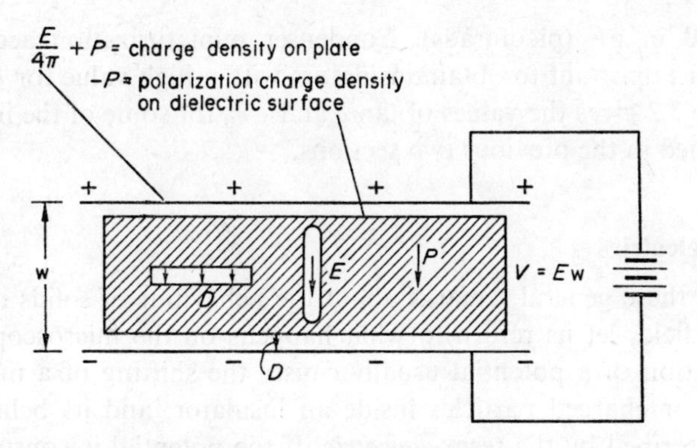

FIG. 5.7. Definitions of D and E (based on Kittel, 1953).

ϵ of the medium filling the space between the two plane electrodes. The value ϵ_0 relating to the vacuum plays a special role, and the properties of an insulating material are evaluated in relation to it by means of the relative constant ϵ/ϵ_0. The capacity C intervening in the expression of capacitance is approximately equal to $\epsilon S/11.3w$, where S is the surface area of the plates in square centimetres, ϵ is the dielectric constant in c.g.s. units, and w is the thickness in centimetres. The capacity is then

Table 5.2

Material	tan δ	ϵ/ϵ_0
Hypalon 100 Hz	0.014	6.19
1 kHz	0.020	6.04
100 kHz	0.098	5.10
PVC 100 Hz	0.082	7.0
1 kHz	0.12	5.9
100 kHz	0.13	4.0
Teflon	0.0001	2.2
Polystyrene	0.0001	2.6
Polyethylene	0.0001	2.2
Mylar	0.001	2.8
Low-voltage porcelain	0.01–0.02	6–7
High-voltage porcelain	0.006–0.01	6–7
Alumina porcelain	0.001–0.002	8–9
Steatite	0.0008–0.0035	5.5–7.5
Forsterite	0.006–0.002	8–9

obtained in pF (picofarads). Condenser miniaturization accordingly makes it important to obtain insulators with a high value for ϵ.

Table 5.2 gives the values of tan δ and ϵ/ϵ_0 for some of the insulators mentioned in the previous two sections.

5.4. Dielectrics

After these general remarks about the behaviour of solids in an a.c. electric field, let us return to what happens on the microscopic scale. Application of a potential usually causes the shifting of a number of dipoles or charged particles inside an insulator, and its behaviour is then described by the term *dielectric*. If the potential is constant, only the overall result can be observed. If it is alternating, more detailed analysis is possible, provided that the local field E_0 at each point of the solid and the number N of dipoles or particles per unit of volume around this point are also brought in. It is written that polarization is the sum of local terms, each of which is proportional to the product of the preceding quantities

$$P = \Sigma \alpha_i N_i E_{0i}.$$

Proportionality coefficients α_i are called *polarizabilities*.

Depending on the frequency of the a.c. potential, various contributions can then be distinguished, as shown in diagrammatical form in Fig. 5.8. The elementary dipoles formed inside a heteroatomic molecule have the greatest inertia, and are accordingly the first no longer to be orientated in the field when radio frequencies are exceeded. It is then the turn of the atoms themselves, carrying the effective charge e^* (cf. § 2.5), no longer to follow when infrared light frequencies are attained. Finally come the electrons, which possess the lowest inertia, when the frequencies of ultraviolet light are reached. In solids, such a general schema requires the following reservations. First, dipoles, connected to the lattice, can seldom become orientated in the field. Second, the presence of uncrossable interfaces introduces an additional term because of the space charges produced in their neighbourhood.

But electrostatic actions cannot be separated from mechanical actions since all the atoms in the solid are connected with one another. One can understand, then, that polarization is accompanied by a

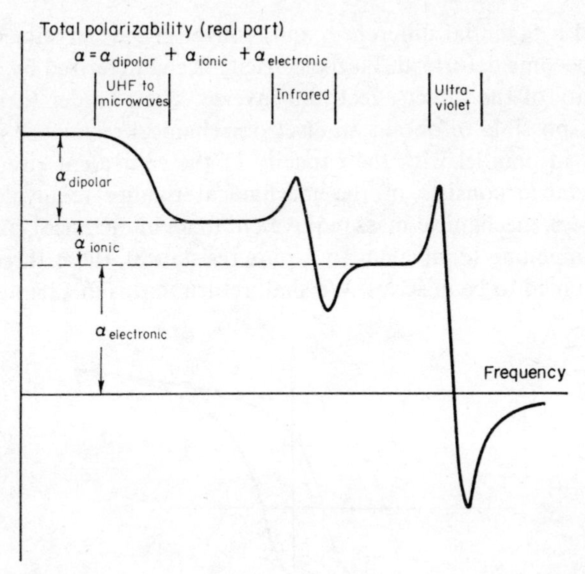

FIG. 5.8. Frequency dependence of the several contributions to polarizability (based on Kittel, 1953).

change in the positions of equilibrium of atoms and the dimensions of the sample being studied: this is *electrostriction*. If the arrangement of atoms comprises a centre of symmetry, which happens in thirty-two classes of crystalline symmetry, the phenomenon is proportional to E^2, and accordingly independent of the sign of E. If the arrangement does not contain any centre of symmetry, the phenomenon is proportional to E and becomes inverted with the sign of E. Although there are twenty-one classes without such a centre, there is an exception, and only twenty thus present reversible electrostriction, known as the *piezoelectric* effect. Of the twenty preceding classes, half possess at least one polar axis, and solids crystallizing in these ten classes accordingly show spontaneous polarization, varying with temperature, in the direction of this axis, and even in the absence of any applied field. This is the *pyroelectric* effect.

The piezoelectric effect was discovered in 1880 by the Curie brothers in certain asymmetrical crystals such as quartz, tourmaline, and Seignette salts. When compressed in given directions, these materials developed a potential difference, and, conversely, when subjected to a voltage, become deformed. Piezoelectricity is characterized by complete reversibility of the direct effect and inverse effect. Under these conditions it is possible to obtain an electromechanical resonance when the reactance in parallel with the capacity of the equivalent circuit is nil. This reactance consists of the mechanical rigidity (equivalent to a capacitance), mechanical mass (equivalent to an inductance), and mechanical dampening (equivalent to a pure resistance), these three factors being assumed to be in series. We shall return to this in Chapter 10.

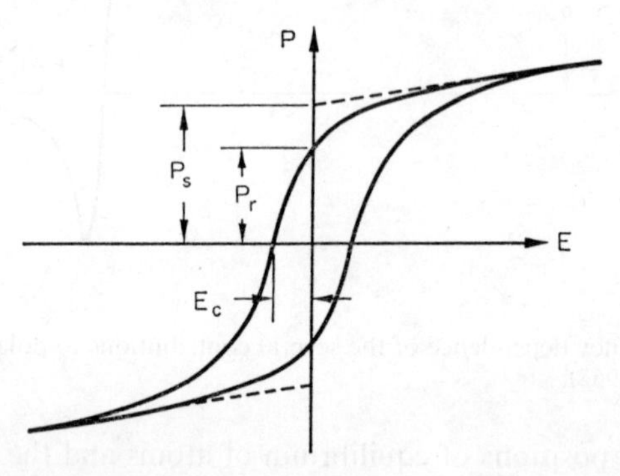

FIG. 5.9. Ferroelectric hysteresis cycle with saturation (P_s) and remanent (P_r) polarization, and coercive field (E_c).

In certain compounds crystallizing in one of the ten pyroelectric classes of symmetry, the crystal structure is relatively unstable, and at high temperature gives way to a higher symmetry structure, often non-pyroelectric. It can also happen that the spontaneous polarization is reversed, in the less symmetrical structure, by applying a sufficiently intense opposing electric field. Polarization then shows a hysteresis cycle under the effect of an a.c. field, and this phenomenon disappears more or less suddenly at the phase-change temperature (Fig. 5.9).

The superficial but obvious analogy with the hysteresis of magnetization of ferromagnetics, which disappears at the Curie point, giving way to paramagnetism, explains the names *ferroelectric* and paraelectric given to the two temperature ranges, and the recurrence of the expression "Curie point". Although domains, within which electric dipoles all have the same orientation, exist in them as in magnetic crystals, this analogy cannot be taken very far. For instance, antiferroelectricity results from antiparallel orientation of domains—and not from adjacent dipoles, as in antiferromagnetism. But it is above all essential to note that the order of dipoles here results directly from their interaction energy W which is very high (W/k varies from 100 to 5000°), while the order of magnetic moments results from electronic interactions in the bonding orbital functions, their interaction energy always remaining very low (W/k around 1°).

5.5. Ferroelectrics

In a diagonal cubic lattice, in other words in which polarization occurs along the diagonal of the cube, the local field is given by the equation

$$E_0 = E + 4\pi P/3 = E + 4\pi \alpha N E_0/3,$$

whence

$$E_0 = E/(1 - 4\pi \alpha N/3).$$

It can be seen that the local field can have very high values, quite out of proportion to the external field applied, if α approaches the "catastrophic" value $\frac{3}{4}\pi N$. This is called "suprapolarizability", and one can see that polarization can be stable in the absence of any applied field. This is what happens in a ferroelectric. The first theory of the phenomenon was produced by Devonshire in 1949, and a new approach proposed 10 years later by Cochran, linking ferroelectric phenomena to the theory of elasticity in crystals.

Microscopic mechanisms responsible for spontaneous polarization still remain largely unknown, however. If the existence of interactions tending to restore a paraelectric structure with high symmetry is proved by the stability of this structure at high temperature, "polarizing" interactions are certainly not confined to lowering the electrostatic energy of the ionic lattice by polarization. Megaw has suggested bringing in the covalent aspect of bonds, the directional character of which would tend to modify the bonding angles determined by the simple assembly of ions. This incomplete knowledge of elementary mechanisms, combined with the diversity of physical phenomena associated with ferroelectricity, makes it difficult to define experimental criteria of the state. There is general agreement, however, on direct observation of a change in the arrangement of domains during reversal of the field, possibly observation of polarization saturation in high fields, all combined with presumptions constituted by the existence of a phase change, if the low-symmetry structure belongs to one of the ten pyroelectric classes.

More than 100 ferroelectric compounds are at present known, without taking account of the solids solutions they often give with one another or other compounds. Their great variety of crystallographic structures and properties rules out any overrigorous classification. The best such classification was given by Merz in 1962. Although it refers very summarily to their mechanical qualities ("soft" and "hard" ferroelectrics), it turns out to coincide approximately with a distinction based on more fundamental criteria, such as the nature of the chemical bonds and of the phase change at the Curie point, and the importance of experimental effects.

Soft ferroelectrics comprise compounds containing hydrogen bonds. The Rochelle salt $NaKC_4H_4O_6$, $4H_2O$, in which ferroelectricity was discovered by Valasek in 1921, and its deuterium homologue were used as a basis for work during the first decade (1925–35). The phosphate KH_2PO_4, discovered by Busch and Scherrer in 1935, as well as the isomorphous compounds XH_2YO_4, provided the basis for research during the second decade (1935–45). More recently, the discovery by Holden *et al.* in 1955 of the guanidine aluminium sulphate with $6H_2O$ (GASH), by Matthias *et al.* of glycocolle sulphate (TGS), and of others such as $(NH_4)_2SO_4$ and $(NH_4)_2BeF_4$, provided the basis for research during the fourth decade (1955–65). Electric dipoles result from the deformation of the ionic groups SO_4^{2-}, BeF_4^{2-}, PO_4^{2-}, etc., in which the central ion is shifted from its equilibrium position, and align spontaneously by the co-operative action of the hydrogen bonds. The phase change at the Curie point consists of an order–disorder transition, and the entropy variation is high, around 1 cal/mole degree. Their Curie constant is low, around 10^3 c.g.s.

Hard ferroelectrics comprise double oxides in which the crystallographic structures contain a lattice of oxygen octahedrons: $BaTiO_3$, discovered by Wul in 1945, then $KNbO_3$ and $KTaO_3$ discovered by

Matthias, $PbTiO_3$ by Smolenskii, and WO_3 by Magascuva, crystallized in the perovskite structure, modified in the case of WO_3; $Cd_2Nb_2O_7$ discovered by Cook and Jaffe, then $Sr_2Ta_2O_7$ in the pyrochlorine structure or neighbouring structures; $PbNb_2O_6$ and $Ba(Nb_{1.5}Zr_{0.25})O_{5.25}$ in the structure of tungsten bronzes; $PbBi_2Nb_2O_9$ in a layer structure; $LiNbO_3$ in the structure of pseudo-ilmenite. Research work during the third decade (1945–55) was based on these. The central Ti, V, Nb, or Ta ion, with the electronic configuration of a rare gas (Smolenskii's rule), is displaced there from its equilibrium position. Coupling between dipoles is provided by oxygens shared by seven octahedrons. The phase change at the Curie point results from the displacement of certain ions in the elementary mesh, and is accompanied by only a slight variation in entropy. Their Curie constant is high, around 10^5 c.g.s.

A few compounds do not fit easily into this summary classification: certain alums, NO_2Na and KNO_3 which, apart from the absence of hydrogen bonds, would be closer to soft ferroelectrics, thiourea $(NH_2)_2CS$ which has the unique distinction of being a molecular crystal, and the compounds SbSI and $Fe_{1-x}S$ which are semiconductors.

Table 5.3 below gives the Curie points and saturation polarizations of some ferroelectric materials. The values of the dielectric constant are considerable, amounting to 100,000 for certain materials close to the Curie point. Figure 5.10 illustrates the variation in the dielectric constant of barium titanate during its successive phase changes.

Table 5.3

Crystal	T_C (°C)	P_s (e.s.u.)
$NaKC_4H_4O_6, 4H_2O$	24	800
$LiNH_4C_4H_4O_6, H_2O$	−167	630
KH_2PO_4	−150	16,000
KD_2PO_4	−60	27,000
$BaTiO_3$	118	48,000
$KNbO_3$	435	
$NaNbO_3$	640	
$LiTaO_3$		70,000 (425°C)

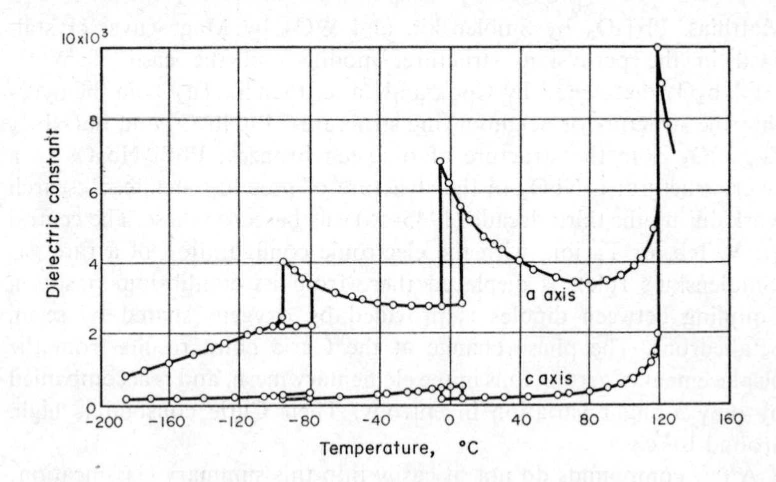

FIG. 5.10. Variation in the initial dielectric constant of barium titanate parallel (*c* axis) and perpendicular (*a* axis) to the direction of the spontaneous polarization (at room temperature) in relation to temperature (based on Merz, 1949). Low temperatures: trigonal; −90–0°C: orthorhombic; 0–120°C: quadratic; over 120°C: cubic structure.

References

BUSSE, W. F. (1954) *Dielectric Materials and Applications* (ed. A. R. von Hippel), p. 168, Wiley, New York.

KITTEL, C. (1953) *Introduction to Solid State Physics*, Wiley, New York; and Chapman, London.

MACKENZIE, J. D. (1964) *Modern Aspects of the Vitreous State* **3,** 126 (ed. J. D. Mackenzie), Butterworths, London.

MERZ, W. J. (1949) *Phys. Rev.* **76,** 1221.

NÉEL, J. and DUPUIS, P. (1972) *Séminaires de chimie de l'état solide* (ed. J. P. Suchet), **5,** 91, Masson, Paris.

SALMANG, H. (1951) *Die Keramik*, Springer, Berlin.

SUCHET, J. P. (1962) *Chimie physique des semiconducteurs*, Dunod, Paris (English translation, van Nostrand, London, 1965).

SUCHET, J. P. (1969) *C.R. Acad. Sci. Paris* **268 C,** 891.

THURNAUER, H. (1954) *Dielectric Materials and Applications* (ed. A. R. von Hippel), p. 180, Wiley, New York.

TOPTCHIEV, A. V., GEIDERIX, M. A., DAVIDOV, D. E., KARGIN, V. A., KRENTSEL, B. A., KOUSTANOVICH, I. M. and POLAK, L. C. (1959) *Dokl. Akad. Nauk SSSR* **128,** 312.

VON HIPPEL, A. R. (1954) *Dielectric Materials and Applications* (ed. A. R. von Hippel), p. 16, Wiley, New York.

Part II

Possible Applications

Chapter 6

Conductors

6.1. Electricity lines

The main application of the electrical conduction of metals is the transmission of current over long distances by wires or cables, usually suspended on pylons above the ground. The mechanical properties of the metal employed are as important in this respect as its electrical properties, as is illustrated by a brief reference to the mechanical calculation of such lines. The equation of equilibrium of a wire, held at both ends and subject only to the effects of gravity, is expressed simply by means of the expression known as "catenary curve" if the origin of the axes is taken at its lowest point:

$$y = K \operatorname{ch}(x/K).$$

In practice, the error made will usually be slight, at any rate for small spans, if the hyperbolic cosinus is replaced by the first two terms of its development in series, in other words if the catenary curve is replaced by the parabola at a tangent to the origin (Ailleret, 1945–6):

$$y = x^2/2K.$$

The parameter K represents the common radius of curvature of the catenary curve and of the parabola at their lowest point (origin of the axes), but it is also the ratio T/P of the horizontal component of the constant tension of the wire between two supports to the force applied per unit of length, namely the resultant of the weight and possibly wind. If the effect of wind is ignored, the parameter K is simply the length of wire—regardless of cross-sectional area—such that, suspended vertically, its upper part is subject to tension T. It is usually around 0.2 L, where L is the length of wire the weight of which is sufficient to break it, the possible expression of mechanical strength (approximately 5–10 km). In a mechanical calculation of lines, technicians use a "change of state equation" in which a certain expression of the tension of the wire, force applied per unit of length, modulus of elasticity, coefficient of expansion, temperature, length, and cross-sectional area of the wire constitutes an invariant. They can then calculate the effect of wind and temperature on the wire, and deduce the optimum distance apart of the pylons, taking topographical relief into account.

The wire conveying electricity to the user has a certain resistivity ρ, and is accordingly subject to temperature rise through the Joule effect, resulting in a voltage drop $\rho Il/s$, where I is the current in the line, l its length and s its cross-sectional area. For a given intensity, these losses increase in inverse proportion to the cross-sectional area, but the cost of the line obviously increases in direct relation to the cross-sectional area. There is thus an economic optimum for this area. Figure 6.1 shows, in diagrammatical form, the result of lengthy calculations, involving various factors, and reveals the existence of a minimum cost

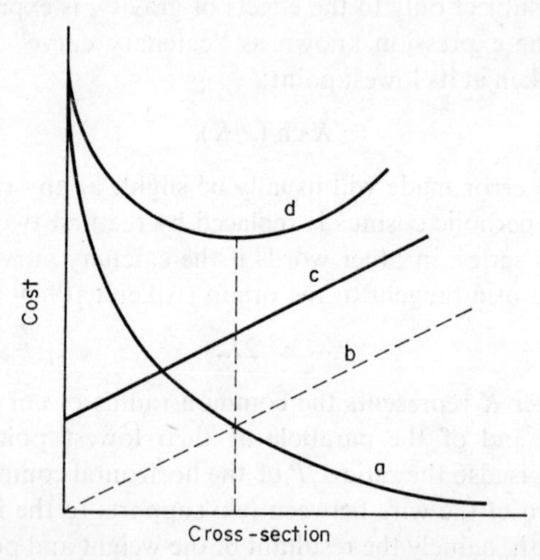

FIG. 6.1. Variation in the cost of an electric line in relation to its section: (a) capitalized value of the Joule effectlosses, (b) fraction of the cost of the line which is proportional to the section, (c) cost of the line, (d) total cost (based on Ailleret, 1945–6).

for a certain cross-sectional area. In general, the slope of the curve *d* close to the minimum is fairly slight, allowing some adjustments. Account also has to be taken of the type of utilization of the line, whether always fully loaded or with widely varying loads. In certain very special cases (underground lines), the slight thermal dissipation means that major importance has to be attached to the Joule-effect temperature rise, to determine the lower limitation of cross-sectional area.

These considerations mean that a choice has to be made among the highly conducting metals mentioned in Chapter 1. Gold and silver are too expensive. There remain copper and aluminium. The breaking strength and elasticity modulus of aluminium are too low to allow it be used alone, so that it is combined with steel in cables, that usually comprise seven strands of steel to thirty strands of aluminium, all of the same diameter. Table 6.1 shows the comparative characteristics of aluminium–steel and copper cables, as well as of all-aluminium or all-steel cables. The product of resistivity × density, on which depends the weight of metal needed to obtain the same conductance, is 12.3 for aluminium–steel and 15.3 for copper, giving an advantage of approximately 20% for aluminium–steel. Furthermore, aluminium is produced in the European Community, although copper is not. The respective prices of the two metals have varied considerably and will continue to do so, but for the same or similar price, aluminium–steel appears to offer an advantage. Sagging, notably, is slighter, allowing spans to be increased or the heights of pylons reduced. It might also be pointed out that earthing cables, which protect very high-voltage lines against over-

Table 6.1. Based on Ailleret, 1945–6

	Aluminium strands	Steel strands	Aluminium–steel cable	Copper cable
Resistivity 20°C (μΩ cm)	2.8	19	3.3	1.76
Density	2.7	7.8	3.75	8.9
Breaking strength (kg/mm²)	17	120	32	40
Elasticity modulus	5000	20,000	8000	10,500
Expansion coefficient $\times 10^{-6}$	23	11.5	18	17

voltages with an atmospheric origin (and which are attached to the tops of pylons), are not traversed by any high current, and are frequently constructed of steel, which is much cheaper. However, such cables need some supervision because of their tendency to oxidation. Finally, lead is used, in the form of a press-moulded pipe, to keep underground cables watertight, after insulation.

6.2. Telecommunications by wire

Telegraphy is literally the art of corresponding at a distance. Beacons lit on hilltops were succeeded by more elaborate optical signals, and the Chappe telegraph, which appeared in 1791, was used successfully throughout the world for nearly a century. Telegraphy came into its own in the modern world, however, with electrical signals, propagated at great distances by metallic conductors. The first decisive test, carried out in 1838 by the American Morse, involved a distance of some 5 km. A few years later began the epic story of the laying of the first underwater cables, and 30 years afterwards all the continents were linked with one another. Meanwhile, transmission and reception of signals by hand had been replaced by Hughes' and Baudot's instruments, which have now been superseded by teleprinters, providing automatic transmission of up to 400 signals a minute.

In contrast to telegraphy, telephony really came into existence only at the electrical stage. The Englishman Hooke's string telephone (1667) and the Frenchman Dom Gauthey's acoustic tube (1680) are of only anecdotal interest. The electric telephone appeared in 1876 when the American Bell, combining partial discoveries by several other inventors, presented a well-designed apparatus at the Philadelphia Exhibition. Within a few years it had spread throughout the world. But it soon raised much more complex problems than the telegraph: telephone exchanges sorting calls and directing them to other exchanges, low-distortion circuits, and the space needed for transmission wires. We shall consider these last two points at greater length.

In telegraphy, as in telephony, the single copper wire very soon gave way to cables consisting of several wires, the structure of which became increasingly complicated. These are telecommunication cables, the component materials and technology of which have developed considerably in the last 100 years. This cable can be defined as a set of insulated conductors, held in a hermetic covering, which may be protected by armouring or sheathing. No outside electrical disturbance can exert any inconvenient influence inside, and no inside electric field must radiate to any significant extent outside (Bélus, 1961). A distinction is made between cables with symmetrical circuits in which the two conducting wires of each telephone circuit perform comparable roles, and cables with coaxial pairs in which one of the conductors is a wire placed inside the second, which is cylindrical in shape.

Cables with symmetrical circuits contain copper wires varying in diameter from 0.4 to 1 mm for short distances, but this diameter is standardized to 0.9 mm for long distances. Insulation has usually been provided by means of strips of paper, which are increasingly being superseded by the technique of extruding a polyethylene tube round the conducting wire. There are two types of wiring. In the first, two insulated wires twisted round each other form a *pair*, and two such pairs twisted in the same way round each other form a *quad* (Fig. 6.2). The transmission capacity is 50% greater than for wiring made up of single pairs through the use of a combined circuit (one wire in one pair and one wire in the other): hence the name "quad with combinable pairs". In the second type, four insulated, equidistant wires are twisted in such a way that their centres always form the apices of a square: hence the name "star quad". This arrangement, which is more compact than single-pair wiring, gives a transmission capacity about 25% greater for a given cable diameter.

The need to transmit the largest possible number of simultaneous telegraphic signals on the same cable has led to the use of alternating currents of different frequencies which are separated on arrival. In telephony, the human voice already covers a certain frequency range, so that the number of simultaneous communications allowed by a similar process is smaller. In both cases, cables are used to transmit alternating currents with the highest possible frequencies. In addition, the reciprocal capacity of two wires in a cable has to be taken into account: it can vary from 20 to 50 nF (nanofarads) per km, depending on types of wiring and insulation, and it constitutes the main cause of fading and distortion of signals over long distances. The first remedy used was to insert, at points on the wires, coils with a magnetic core

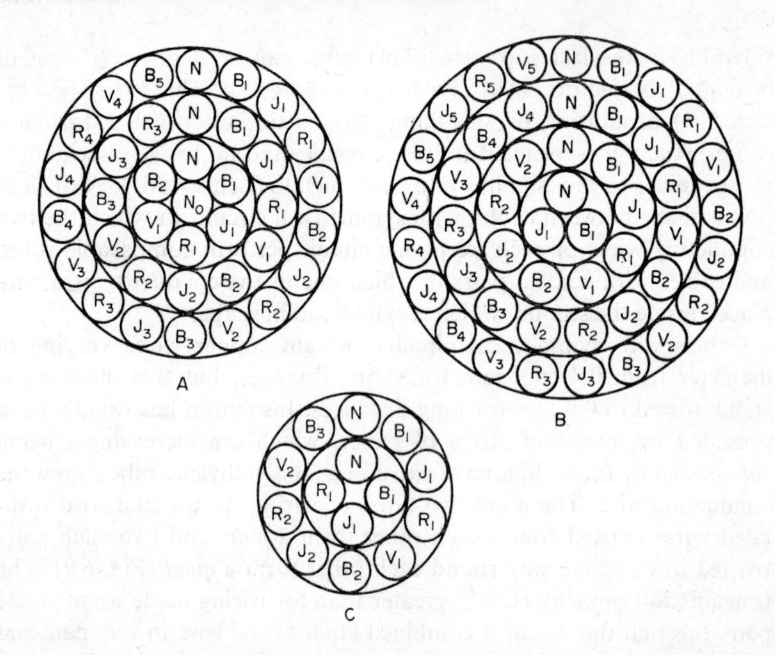

FIG. 6.2. Sections of homogeneous standardized cables with 37 (*A*), 48 (*B*), and 14 (*C*) quads with combinable pairs (based on Bélus, 1961). N, B, J, R, and V refer to the different colours of the quads.

without losses (ferrites), so as to reduce the (negative) reactance to a smaller absolute value (cf. § 5.3). This produces a cut-out frequency, however, which runs counter to the aim sought, and losses are increased. For this reason, the present tendency is to reduce the number and value of such coils, and to amplify the signals more frequently by equipping cables with *repeaters*, which may be transistorized and remote-powered.

It was the need to transmit a much wider frequency range, notably for television pictures, that led to cables with coaxial pairs, at present standardized at 1.2/4.4 and 2.6/9.5 mm; the latter transmit a wide frequency band of 4 MHz with an amplification distance of 9 km. The two conductors are made of copper, and the tube consists of a stapled tape with an internal diameter of 9.5 mm. Two steel ribbons, wound in a spiral round the pair, provide a binding as well as a hermetic covering

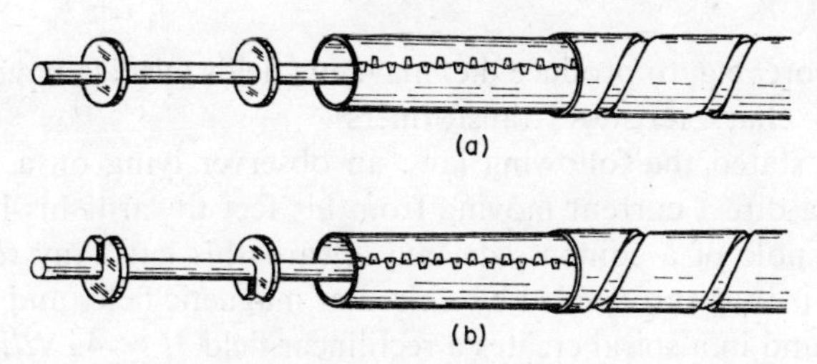

FIG. 6.3. Coaxial standardized 2.6/9.5 mm pair, insulated by moulded (*a*) or crimped (*b*) discs (based on Bélus, 1961).

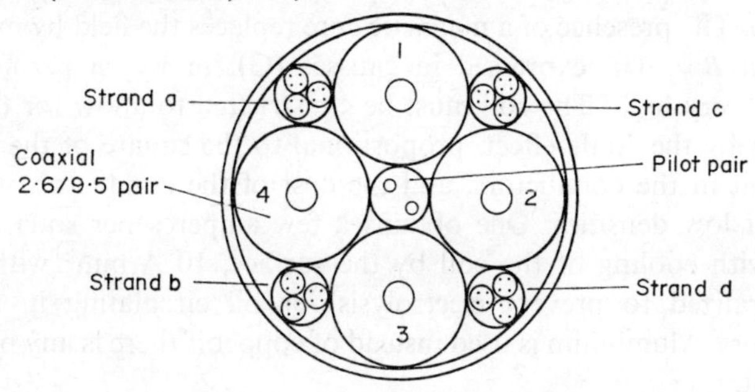

FIG. 6.4. Section of the composite standardized cable with 4 coaxial 2.6/9.5 mm pairs, 12 0.6 mm star quads, and one pilot 0.9 mm pair insulated with polyethylene (based on Bélus, 1961).

and electromagnetic screen (Fig. 6.3). Insulation is provided by 2 mm polyethylene discs, 28 mm apart. The average relative dielectric constant of the dielectric thus obtained is low, at around 1.09, giving a capacity of 47 nF/km. The d.c. resistance of the inside wire is 3.1 Ω cm at 10°C. Very often, a single composite cable comprises coaxial pairs and quads (Fig. 6.4).

The descriptions given above apply to all sorts of cables; for underwater cables, however, special protection is provided against corrosion by sheets of lead and gutta-percha, while for overhead cables, steel armouring enables them to withstand the tensile stresses resulting from their weight and from wind.

6.3. Magnetic coils

Conductor metals are frequently used to make coils with a permeable magnetic core, e.g. to produce d.c. magnetic fields (electromagnets) and change a.c. characteristics (transformers).

Ampère stated the following law: an observer lying on a wire traversed by a direct current moving from his feet towards his head sees the North pole of a compass deviate towards his left. Any rectilinear conductor is thus surrounded by a circular magnetic field, and any conductor wound in a spiral creates a rectilinear field $H = 4\pi NI/l$, where I is the current, N the number of spirals, and l the length of the solenoid. One obtains approximately 1.25 Oersteds for 1 ampère (A), 1 spiral, and 1 cm. The presence of a magnetic core replaces the field by magnetic induction $B = \mu H$, expressed in gausses (G), in which permeability μ can be very high. The coil must be constructed to allow for the heat given off by the Joule effect, proportional to the square of the density of current in the conductors, and the cost of the conductor, which is higher at low densities. One obtains a few ampères per square millimetre, with cooling of the coil by the surface, 10 A/mm^2 with water (demineralized to prevent electrolysis) or oil circulating in tubular conductors. Aluminium is used instead of copper if there is any problem of weight.

The coil is generally constructed round a magnetic carcass or, in electronics, round M- or EI-shaped iron–nickel sections or ferrite shells. Figure 6.5 illustrates the main shapes. For electromagnets, wrought-iron or laminated sheet-iron are used up to inductions of 30–40 kG. Permeability has to be higher for transformers, and losses through eddy currents (induced in the mass of the core by the variable field) smaller: this has led to the use of laminated iron–nickel conductor alloys containing 75% nickel (permeabilities of 10,000–100,000, saturation induction of 6–9 kG), or, at higher frequencies, sintered semiconductor ferrites of nickel–zinc or manganese–zinc (permeabilities of 100–2000, saturation inductions of 2–3 kG). The aperture in the sheet or carcass must hold the conductors and all the materials needed to insulate them. Its surface area is thus considerably greater than the cross-section of the metal. Tables in technical works show the outside diameters normally employed by manufacturers after insulation by enamel and silk for a given diameter of bare conductor. In each case formulae and diagrams allow the number of spirals per square centimetre to be calculated.

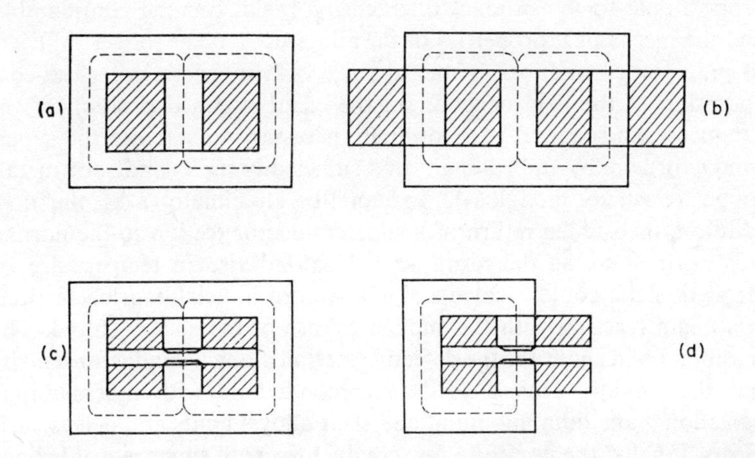

FIG. 6.5. Main types of magnetic circuits: (*a*), (*b*) transformers; (*c*), (*d*) electromagnets.

As for a condenser, consideration of a coil includes a quality or overvoltage factor Q, defined as the ratio of its reactance to its resistance (assumed to be in series) and a tangent of the angle of loss $\tan \delta = 1/Q$ (cf. § 5.3). These quantities are used mainly at high frequencies, notably with ferrite-core coils, and $\tan \delta$ is then the sum of two terms, one resulting from the ohmic resistance of the wire and the other from eddy currents in the core.

The production of d.c. magnetic fields is used for instance in electromagnets designed to handle scrap iron, but scientific research laboratories also make wide use of them. For high inductions, above 40 kG, the magnetic core disappears, since the cost of energy needed for its saturation would be prohibitive. In addition, ordinary conductors are beginning to be replaced by superconductors, of which this is the first major application. Intense inductions, of around 100 kG, can then be obtained in electromagnets of reduced size, since there is no further loss through the Joule effect, and since current densities of 100 A/mm^2 are possible.

However, this result is obtained only at the cost of serious technological complications. The forces to which conductors are subject, proportional to the product of intensity × field, become considerable, and the mechanical properties of the alloys used are no longer sufficient to prevent deformation. In the absence of a magnetic core, the configuration of the field depends on the clearly defined position of the conductors in space. It is accordingly necessary to stabilize the superconductor wire by embedding it in a sufficiently rigid conductor metal: copper (elasticity modulus 11 kg/mm^2) or aluminium (3 kg/mm^2). In addition, the sudden return of a superconductor section to the normal conductor state, as the result of a localized rise in temperature or magnetic field, could involve a similar return in neighbouring sections in a chain reaction with the sudden release of the energy stored. The possibility of a shunt for the defective section along a conductor matrix, and the use of "second-class" superconductors with a less-abrupt transition (vanadium, niobium, and their alloys) limits and delays such effects. Finally, the use of hollow conductors, with supercritical helium circulating inside them under pressure, ensures more uniform cooling of the superconductor material (Fig. 6.6).

Two recent cases may be mentioned. NASA in the United States uses metal strips covered with a thin layer of Nb_3Sn, and then a thin layer of silver (total thickness approximately 0.1 mm), alternating with strips of Mylar-insulated copper, to obtain 140 kG in a net diameter of 150 mm (Schrader and Thompson, 1969). In the other case, CEA in France uses NbTi wires in an aluminium matrix to obtain 66 kG in a net diameter of 135 mm (Adesio and Bronca, 1970).

The need to modify the characteristics of alternating currents results primarily from the conveyance of electrical energy. Let us assume that 50 kW d.c. has to be transmitted 20 km without losing more than 10%. If 500 V × 100 A is chosen, the loss of 5 kW through the Joule effect would require a total line resistance of 0.5 Ω which would involve a copper wire 42 mm in diameter and weighing 425 tons. By opting for 10,000 V × 5 A, on the other hand, the corresponding figures are 200 Ω, 22/10 mm, and 1 ton. For a given power loss, the cross-sectional area of wires, and accordingly their weight and cost, are inversely proportional to the square of the voltage (Carton, 1948). This conclusion applies qualitatively to transmission at present generalized in a.c., and up to 225 or even 380 kV are used: hence the need for numerous step-

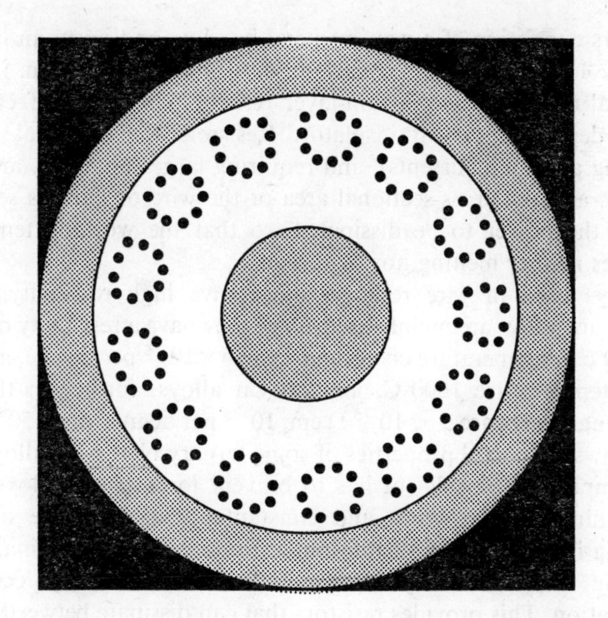

FIG. 6.6. Hollow conductor (diameter 4 mm) with 130 wires (diameter 150 μm) of superconducting Nb–Ti alloy in an insulated copper matrix (diameter 10 mm) (based on Bidault and Dosdat, 1970).

up and step-down voltage transformers. Furthermore, low-power transformers have very many uses in electronics.

The operating principle of a transformer is simple: a primary winding of N spirals creates a magnetomotive force NI in a closed magnetic circuit, and the same force creates an alternating current I' in a secondary winding of N' spirals, such that $NI = N'I'$, so that $V/V' = N/N'$.

6.4. Resistors and couples

Resistors are circuit components in which the current passing through them is proportional to the voltage applied to their terminals. A distinction is usually made between fixed resistors, the value of which is shown by a colour-ring code, variable resistors (rheostats, potentiometers), and, less frequently, adjustable resistors. We prefer to make a distinction based on the materials used, so that there are wound resistors consisting of a wire or metal band wound on to an insulating shaft, composite resistors consisting of a mixture of carbon, insulating material and binder, and thin-layer resistors consisting of carbon or metal deposited on an insulator. Resistors for electrical purposes (heating and high currents), and requiring large capacities, are usually wound, and the cross-sectional area of the wire or band is selected to match the power to be dissipated, so that the working temperature involves neither melting nor oxidation.

Alloys used in wire resistors must have high resistivity. Nickel–chromium alloys are mainly used, since they have a resistivity of around 10^{-4} Ω cm, temperature coefficient 1 to 15×10^{-5} per degree, and maximum temperature 1000°C, and Ni–Cu alloys, for which the corresponding figures are 5×10^{-5} Ω cm, 10^{-5} per degree, and 150°C. Table 6.2 shows the useful properties of some ternary Ni–Cr–Fe alloys. When the temperature coefficient has to be very low, special alloys such as manganine $Cu_{84}Ni_4Mn_{12}$ and constantan $Cu_{60}Ni_{40}$ are used. The backing is generally of steatite porcelain or sintered alumina, and the winding is protected by a stove-baked paint, fireproof cement, or vitrification. This provides resistors that can dissipate between approximately 10 W and 1 kW. The appearance of an inductance is prevented by a special winding process.

Certain conductors have special properties. Amorphous carbon, for instance, with a resistivity of 3 to 6×10^{-3} Ω cm, is used to make electrodes for furnaces and electric arcs, and used to be used for in-

Table 6.2

	Resistivity (μΩ cm)	Mean temperature coefficient (0–1000°C)	Curie point (°C)	Limit of use (°C)
$Ni_{12}Cr_{12}Fe_{76}$	74	—	Room temperature	600
$Ni_{36}Cr_{11}Fe_{53}$	100	—	150	700
$Ni_{48}Cr_{22}Fe_{30}$	110	1.2×10^{-4}	None	1000
$Ni_{60}Cr_{15}Fe_{25}$	112	1.3×10^{-4}	100	1100
$Al_5Cr_{32}Fe_{63}$	140	0.6×10^{-4}	380	1200
$Ni_{80}Cr_{20}$	105	0.6×10^{-4}	None	1200

candescent-light filaments, though it has now been replaced by tungsten, which can be raised to higher temperatures. Pure nickel and platinum have a clearly defined resistivity temperature coefficient (although low) and are used in thermometers. Bismuth exists in flat coils of 1000 Ω, the resistance of which increases by about 4% per kG, and which can accordingly be used in gaussmeters. Resistors known as "negative temperature coefficient resistors" or "thermistors" and "non-linear" resistances do not make use of conductors, and their applications are dealt with in Chapter 8 and § 6.5 respectively.

A fine iron-wire resistor placed in a hydrogen atmosphere, and known as a "baretter", plays a rather special role, connected with the magnetic dispersion of iron below its Curie point (cf. § 1.4) and with the high positive temperature coefficient of the resulting resistivity. If the portion of curve $\rho(T)$ below T_C is regarded as approximately exponential, one has

$$\rho = A \exp B/(T_C - T).$$

Since the thermal inertia of the fine wire is low, its heating causes a non-linear variation in the voltage at its terminals, in relation to the current passing through it

$$V = CI^n$$

with $$n = \mathrm{d}(\ln V)/\mathrm{d}(\ln I) = \mathrm{f}(T).$$

Figure 6.7 shows the possible functioning of such a device, on a V–I logarithmic graph. The point T_1' corresponds to n = infinity, in other words current regulation.

Finally, it was pointed out in § 1.4 that the variation in resistivity of an alloy during a phase change may be considerable if the magnetic order is altered. This is the case for $Mn_{2-x}Cr_xSb$ solutions, which crystallize in the quadratic system and show a transition from antiferromagnetism to ferrimagnetism when the temperature rises. Figure 6.8 shows the effect of this phenomenon on resistivity for various values of x. Where $x = 0.1$, resistivity drops suddenly by 27% at 309°K with 1°K hysteresis. Such a component can naturally be used as a thermal detector, as, indeed, can other solids showing similar phenomena (cf. Chapter 9).

When two different metals are in contact, a potential difference

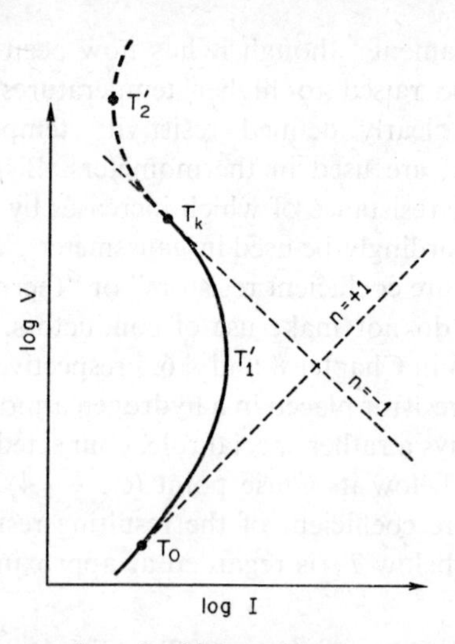

FIG. 6.7. Principle of operation of a high positive temperature coefficient and low thermic inertia resistor (based on Suchet, 1955a).

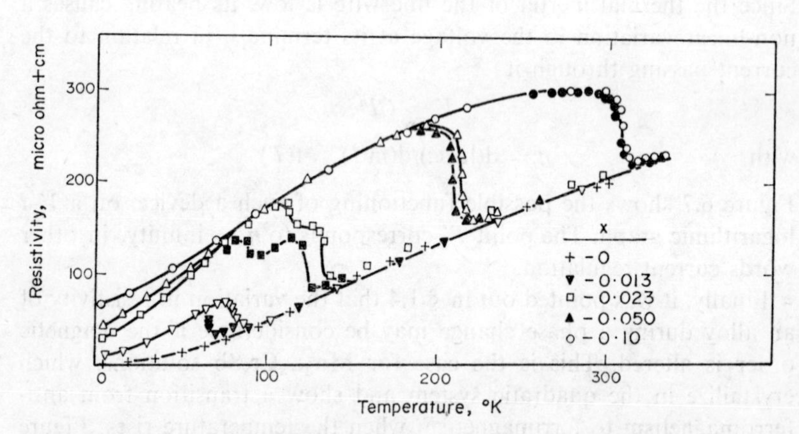

FIG. 6.8. Variation in the resistivity of $Mn_{2-x}Cr_xSb$ solid solutions in relation to temperature (based on Bierstedt, 1963). Black dots show cooling.

exists between them, related to temperature. If the two welds of a bimetallic circuit are not at the same temperature, accordingly, the opposing potential differences will not be equal, and the terminals of any cut-out will still retain a voltage related to the difference in temperature between the two welds. Table 6.3 gives the value of this voltage for some metals, based on platinum as the common partner, and for some couples of alloys in which it is particularly high, with their working temperature limit. The couple platinum/rhodium–platinum is used in all high-temperature furnaces where the exact temperature has to be known. Above 1450°C, optical pyrometry is employed. We shall return to the thermoelectric effects in Chapter 7, in connection with conventional semiconductors, which show such effects much more intensely.

Table 6.3

Fe	21 μV/°C			
Ag	9			
Zn	7.5	Chromel–constantan	62 μV/°C	900°C
Cu	7.4	Iron–constantan	54	900
Sn	4.1	Chromel–alumel	41	1100
Al	3.8	Platinum/rhodium–platinum	6.4	1450
W	1.8	(10% rhodium)		
Ta	1.7			
Ni	−19			

6.5. Contact parts

Electric circuits are generally designed to be open and closed in turn, the frequency varying considerably depending on applications. Breaking devices therefore have to contain contact parts which put up little resistance to the passage of the current and are not damaged by the small arc that occurs when the circuit is broken. The qualities required for the materials used are good electrical and thermal conductibility, low contact resistance, good mechanical properties (high melting point, arc resistance, machinability, and a high level of hardness for pressure contacts, or a low level of hardness, on the other hand, for sliding contacts). Table 6.4 shows the resistivities and hardnesses of some more or less refractory conductor metals, and for the various types of carbon used in sliding contact brushes, notably in rotating electrical machinery (motors and dynamos).

The arc created by the cutting of a circuit results from the surge in voltage connected with the presence of an inductance in the circuit.

Table 6.4

Metals	Hardness (Brinell)	Resistivity ($\mu\Omega$ cm, 25°C)	Brushes carbon	Hardness (shore)	Resistivity ($\mu\Omega$ cm, 25°C)
Ag	35	1.59			
Cu	40	1.67	Hard	70	700
Pt	47	10.6	Soft (graphite)	20	200
Ni	90	6.84			
Mo	250	5.6	Metal–graphite	20	50
W	380	5.5			

The higher this inductance the greater the current, and the slower the break the greater will be the arc. The surge appears between the two components in contact at the precise moment at which they have just separated and are still a very small distance apart, around 0.1 mm for instance. The result is obviously the breakdown of the insulating layer of air and the striking of an oscillating arc, which will stop only when the two parts are sufficiently far apart. Wear on the components making the contact results partly from the removal of metal ions by the electrical discharge and partly from the temperature rise, which may even cause melting. In air, these two phenomena generally combine to form a layer of oxide on the surface of the parts, increasing the resistance of the contacts and speeding up their deterioration. It can be shown that the value of the voltage surge is roughly equal to the product of the currents passing through the circuit before breakage by the total resistance of the circuit (Suchet, 1955b).

The best protections against voltage surges make use of non-linear resistors (cf. § 4.5). These resistors drop as the voltage applied to them increases. A non-linear component, placed in parallel with an inductance, will thus have a high resistance for normal voltage, but will act more or less as a short circuit when the surge occurs. If the inductance is spread through the circuit, the component can be placed in parallel with the contact. Figure 6.9 shows, in diagrammatical form, the variations in voltage at the inductance terminals, namely also at the component terminals, and the variation in its resistance in relation to time. The characteristics of the non-linear component are selected so that its resistance becomes less than that of the layer of air separating the two

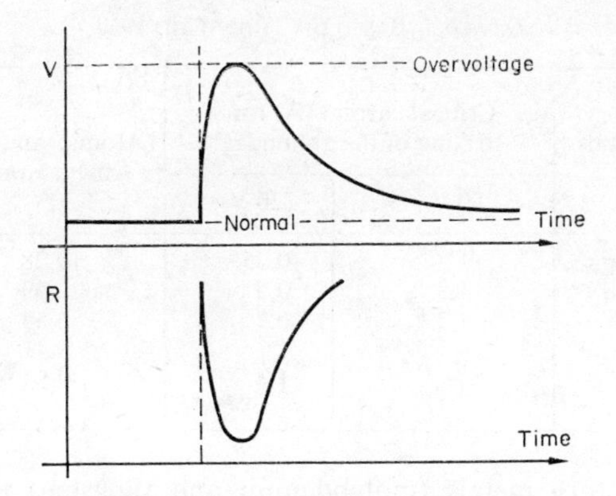

FIG. 6.9. Comparative variations in the voltage at the terminals of a contact and in the resistance of a non-linear element in relation to time.

contact parts, thus preventing an arc from being struck, or extinguishing it quickly. If, for example, a component allowing 3 mA to pass at 250 V and 10 mA at 2000 V is used in a high-inductance circuit with a voltage of 140 V and 10 A current, it will reduce the peak voltage from 10,000 to 1000 V.

However effective methods of protection involving a parallel non-linear resistor, high-speed separation of contact parts and possibly even blowing-out of the arc, the materials used are subject to rigorous conditions, and special compositions and techniques for preparing them are required. This is why, alongside the metals shown in Table 6.4, composite materials, resulting from the powder metallurgy, have also appeared. These "pesudo-alloys" generally combine a fairly insulating refractory metal with a very good conductor metal. The choice of the refractory metal is the more complicated. Table 6.5 illustrates some factors. The intervention of the atomic volume is explained by the importance of the mass of material transferred from one contact part to the other per Coulomb. This is expressed as follows (Holm):

$$m = (UV_A^2/\sqrt{D}) \times C^t$$

where U is the characteristic voltage of the arc, V_A the atomic volume, and D the Brinell hardness.

Table 6.5. BASED ON THIEN-CHI, 1950

Metals	Critical current (A) for striking of the arc under		Atomic volume (cm^3/g atom)
	50 V	220 V	
Ag	1	0.25	10.28
Cu	1.3	0.5	7.09
Ni	1.2	0.7	6.59
Mo	3	1	9.41
W	4	1.4	9.53

Very refractory metals (molybdenum and tungsten) thus provide pseudo-alloys with silver and copper in which the resistivity depends mainly on the quantity of good-conductor metal (approximately 4×10^{-6} Ω cm for 20% silver and 3.5×10^{-6} Ω cm for 40% silver). With nickel, which is less refractory, resistivity is lower. This method can also be used to incorporate a more or less refractory oxide. Silver–cadmium oxide pseudo-alloys, e.g. the one with 9% CdO, appear to offer useful properties, as well as possibilities of sintering, hammering, and drawing similar to those for purely metallic compositions. Resistivity varies evenly from 1.6×10^{-6} Ω cm (for pure silver) to 2.5×10^{-6} Ω cm (for 15% CdO), and they are suitable for contact parts working with low currents.

There remains the case of bimetallic strips, in which the different expansion coefficient of two strips welded to each other causes the composite strip (bimetallic strip) to change its curvature at a given temperature. For certain shapes produced by pressing, this change can be sudden, and can be used to switch a circuit: hence their application in thermal detection.

References

ADESIO, P. and BRONCA, G. (1970) *Annls Mines, Paris* 34, January issue.

AILLERET, P. (1945–6) *Cours d'électrotechnique appliquée*, École Nat. Ponts et Chaussées, Paris.

BÉLUS, R. (1961) *Câbles et Transmissions* **15,** 267.

BIDAULT, M. and DOSDAT, J. (1970) *Annls Mines, Paris* 57, January issue.

BIERSTEDT, P. E. (1963) *Phys. Rev.* **132,** 669.

CARTON, R. (1948) *Transformateurs*, Armand Colin, Paris.
FELDTKELLER, R. *Theorie der Spulen und Übertrager*, Hirzel, Stuttgart (French translation Dunod, Paris, 1969).
HOLM, R. (1956) *Electric Contacts Handbook*, 3rd edn., Springer, Berlin.
JUNG, A. (1953) *Calcul des électro-aimants industriels*, Dunod, Paris.
KEIL, A. (1960) *Werkstoffe für elektrische Kontakte*, Springer, Berlin.
MEADEN, G. T. (1965) *Electrical Resistance of Metals*, Plenum, New York; and Heywood, London, 1966.
SCHRADER, E. R. and THOMPSON, P. A. (1969) NASA-CR-1260, January issue.
SUCHET, J. P. (1955a) *Bull. Soc. fr. Électns Paris* **5**, 274.
SUCHET, J. P. (1955b) *Les Varistances*, Chiron, Paris.
THIEN-CHI, N. (1950) *Annls Radioélectr. Paris* **5**, 339.

Chapter 7

Conventional Semiconductors

7.1. Extrinsic semiconductors

Applications of conventional semiconductor materials always involve impure crystals or, to be more precise, extremely pure crystals into which a very small amount of a clearly defined impurity has been introduced. This operation is called *doping*, and the doped crystal is an *extrinsic* semiconductor. The principle of the electrical action of the impurities is the same in elements and compounds, but is much easier to describe for an element. We will therefore consider the case of an element here.

In covalent crystals in column IV of the Periodic Table—diamond, silicon, germanium, and grey tin—the covalency bond is formed by the pooling of four electrons. Let us assume to begin with that the foreign atom replacing one of the atoms in the lattice is provided with only three valency electrons. Figure 7.1*a* shows a boron atom in a silicon lattice. It is clear that the boron atom cannot supply the four electrons required by the pooling system. The fourth electron will be taken from the normal electronic distribution of the crystal, and will leave a positive hole somewhere, which can move freely within the crystal. It is exactly as if the boron atom had accepted an additional electron, and the chemical formula $Si_{1-x}(B^-p^+)_x$ can be taken, where x represents the fraction of silicon atoms replaced by boron atoms. Atoms of elements such as boron are called *acceptors* (of electrons) in the lattices of elements in column IV. Conductivity varies as follows:

$$\sigma = \sigma_0 \exp(-E_P/kT).$$

Figure 7.1*b* illustrates this situation for an electron energy diagram

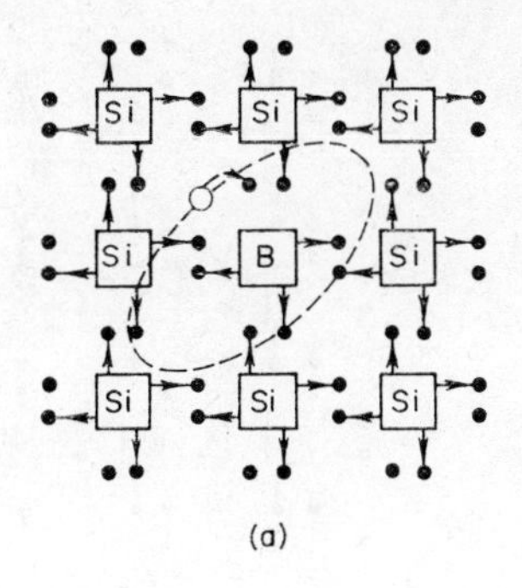

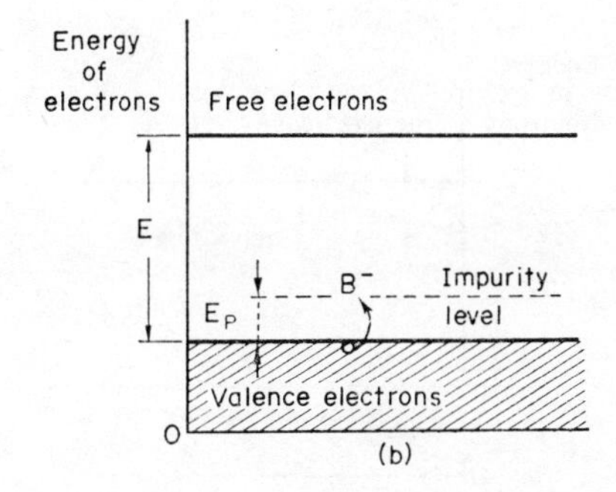

FIG. 7.1. P-type impurity in silicon: bonding schema (*a*) and energy diagram (*b*).

Only the hole that has appeared in the valency electron region can move freely. We thus have semiconduction by holes, and one speaks of a P-type crystal.

Let us now assume that the foreign atom replacing one of the atoms in the lattice is provided with five valency electrons. Figure 7.2*a* shows a phosphorus atom in a silicon network. It is obvious that the fifth electron cannot be integrated in the covalency bonds and will be free to move about in the crystal. It is exactly as if the phosphorus atom had lost an electron, and the chemical formula $Si_{1-x}(P^+e^-)_x$ can be used, where x represents the fraction of silicon atoms replaced by phosphorus atoms. Atoms of elements such as phosphorus are called *donors* (of electrons) in the lattices of elements in column IV. Conductivity varies as follows:

$$\sigma = \sigma_0 \exp(-E_N/kT).$$

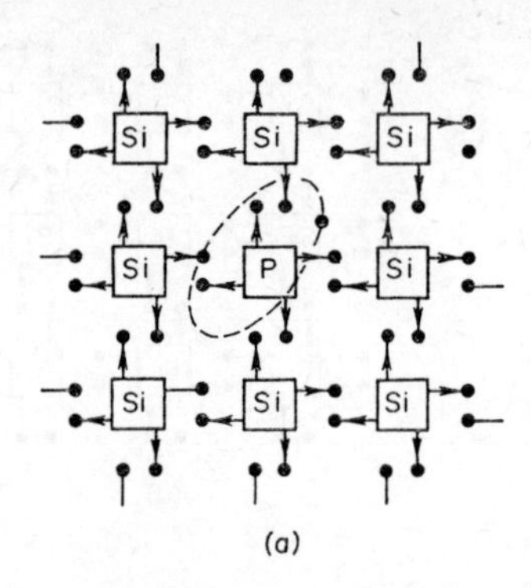

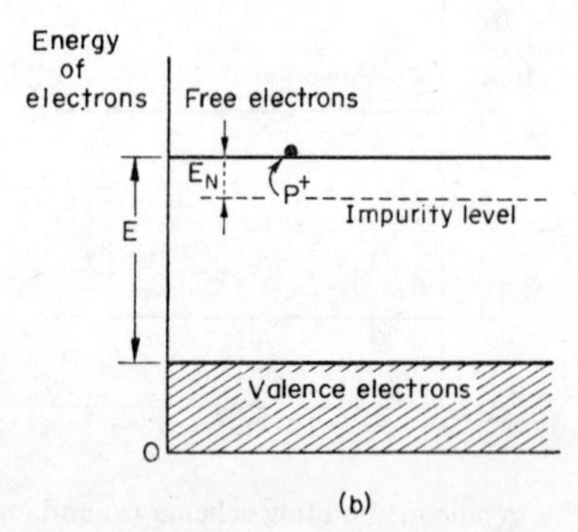

FIG. 7.2. N-type impurity in silicon; bonding schema (*a*) and energy diagram (*b*).

Figure 7.2*b* illustrates this situation for an electron energy diagram. Only the electron that has moved into the free electron region can move freely. We then have semiconduction by electrons, and one speaks of an N-type crystal.

If, instead of an element, a binary semiconductor compound is involved, such as GaAs, the same phenomenon can occur, in relation to the sub-lattice either of the gallium or of the arsenic. At room temperature, for example, an element in column II, such as beryllium, is an acceptor ($E_P = 0.065$ eV), and an element in column IV, such as silicon, is a donor ($E_N = 0.0023$ eV) in relation to the gallium sub-lattice, while an element in this same column IV, such as germanium, is an acceptor ($E_P = 0.035$ eV) and an element in column VI, such as oxygen or sulphur, a donor ($E_N = 0.4$ for oxygen and 0.007 eV for sulphur) in relation to the arsenic sub-lattice. In addition, elements in columns III and V can act as acceptors (Al, $E_p = 0.089$ eV) or donor, but the role is more difficult to predict.

Lattice defects, such as vacancies or interstitial atoms, can perform a similar role to impurities. For example, in an NaCl rock-salt lattice where x chlorine atoms are missing, the ionic formula $Na^+Cl^-_{1-x}\square_x$ shows that x electrons of sodium atoms cannot find any host, and are accordingly in a situation similar to that of the fifth phosphorus electron in the silicon network. The combination of vacancy and electron is called an F centre, and is revealed by a particular absorption band in the spectrum of the salt. In the presence of an electric field, the electron can escape from the vacancy and move in the crystal. Similarly, in the presence of x sodium vacancies, the ionic formula $\square_x Na^+_{1-x}Cl^-$ shows the appearance of x holes on neighbouring chlorine atoms. Such combinations are called V_1 centres, and they too can be dissociated by an electric field, which will displace the holes in the crystal.

The conductivity of such extrinsic semiconductors varies in relation to temperature as follows:

$$\sigma = \sigma_0 \exp(-E_P/kT) \quad \text{or} \quad \sigma_0 \exp(-E_N/kT).$$

The energy E_P (for an acceptor) or E_N (for a donor) is always less than the value of the forbidden energy E. Above a certain temperature, incidentally, all impurity-carrier or defect-carrier combinations are dissociated, and the law of intrinsic semiconductors, which involves E, once again applies.

Finally, passing reference should be made to a new phenomenon which occurs in impure crystals at very low temperatures. Let us assume that a germanium crystal is doped with antimony atoms. This element belongs, like phosphorus, to column V, and should therefore be integrated with the bond lattice in the ionized form Sb^+. At very low temperature, however, only part of these atoms ionize, and the remainder retain the fifth electron in the localized state. A new semiconduction mechanism then appears, involving electronic hops from a neutral antimony atom to an ionized antimony atom (Fig. 7.3). The variation in conductivity in relation to temperature is still exponential, but the Hall coefficient is constant, showing that the number of carriers does

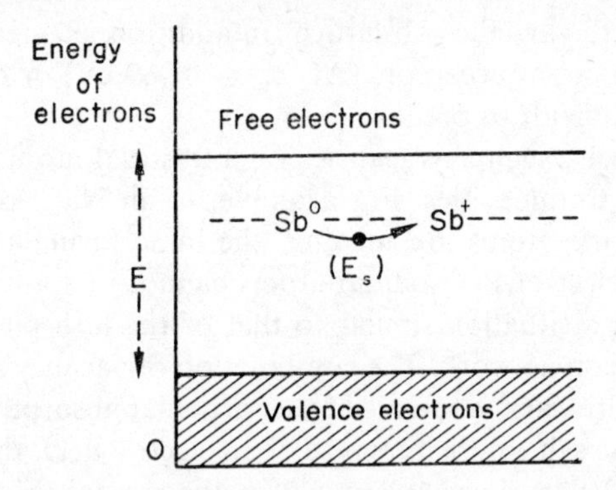

FIG. 7.3. Hopping mechanism schema in the energy diagram of an antimony-doped germanium at low temperatures.

not increase with temperature. Here it is the mobility of the carriers that increases, with the hop becoming easier. Section 3.2 mentioned similar phenomena in semiconductor compounds of transition elements.

7.2. P–N junction: photocell, rectifier

If two regions in one crystal have been doped with impurities that can create conductions by carriers of different types, the boundary between the two regions possesses useful properties. Let us assume that holes, represented by open circles in Fig. 7.4*a* (P region), and that another part of the same crystal contains phosphorus as an impurity, releasing negative electrons, represented by black dots in Fig. 7.4*a* (N region). If these two regions are adjacent, on each side of an ideal junction surface separating boron and phosphorus atoms, the positive holes and negative electrons are not stopped by the junction, and neutralize one another. Only the ionized atoms B^- and P^+ remain. It is this region, without carriers but polarized, that is known as the *P–N junction*.

If the region is illuminated by a light so that $h\nu > E$, pairs of electron–hole carriers are created by a photoelectric effect. Instead of recombining, as they would do anywhere else, these pairs dissociate from each

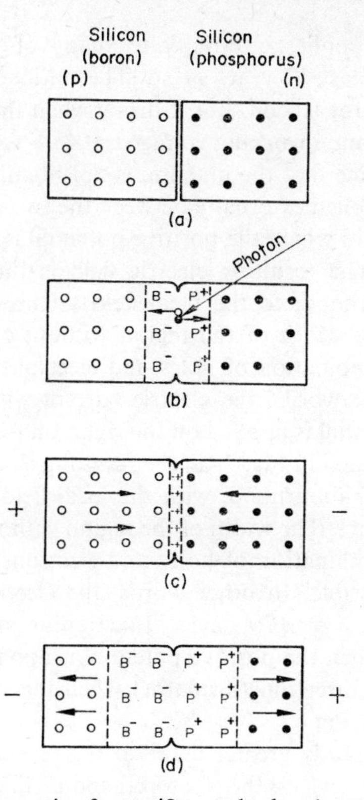

FIG. 7.4. Principles of operation for rectifiers and solar photocells (based on Suchet, 1962).

other under the effect of the electric field produced at the junction by the B^- and P^+ ionized atoms (Fig. 7.4*b*). The electrons are repelled by the B^- negative charges and attracted by the P^+ positive charges and are thus stored in the N region. The holes are subject to opposing forces, and are thus stored in the P region. A potential difference accordingly appears between the P and N regions, resulting in partial compensation of the electric field at the junction, and shrinkage of the region without carriers (Fig. 7.4*c*). Recombination of electrons and holes thus becomes easier, and it can be seen that there is a limit to the storage of electricity in the two regions, and accordingly in the potential difference at the terminals of the device. This is the operating principle of the solar *photocell*, which supplies electricity to many of the space satellites launched over the last 10 years or so. The no-load voltage is around $E/2$, namely 0.6 V for silicon, but it drops when the device is used for charging. The optimum working voltage is 0.45 V for silicon photocells.

Let us now assume that the junction is not illuminated, but a potential difference is applied externally between the two regions. Figure 7.4*c* illustrates the case in which the positive potential is applied on the left, to the P region. The resulting electric field is thus opposed, in the junction neighbourhood, to the local field set up by the B^- and P^+ ionized atoms. The width of the region without carriers will thus be reduced, and recombination of holes and electrons in the two regions facilitated. In other words, the electric current will pass more easily. If the positive potential is applied on the right, on the other hand, to the N region, as illustrated in Fig. 7.4*d*, the resulting electric field combines, in the junction neighbourhood, with the local field created by the B^- and P^+ ionized atoms. The width of the region without carriers will thus increase, and recombination of holes and electrons in the two regions will become impossible. In other words, the electric current will not pass. We thus have a *rectifier* device, functioning in the forward direction (conductor) when the positive potential is applied to the P region, and in the reverse direction (insulator) when the positive potential is applied to the N region.

Let us look in slightly greater detail at the functioning of these first two semiconductor devices, the photocell and rectifier. Photocells make use of very fine sheets of boron-doped monocrystalline silicon, on the surface of which phosphorus is diffused in a markedly higher concentration than the boron. The surface is thus relatively conducting, and a grid-shaped electrode, allowing light through, is sufficient to form the negative pole of the electricity generator. Not all semiconductors, however, are suitable for converting light into electricity. To begin with, the photocell is a selective receiver using only photons of energy $h\nu$ slightly higher than E, resulting in the choice of E at 1.2 to 1.5 eV for the conversion of solar energy. Next, only electron–hole pairs created near the junction or capable of being diffused there very rapidly can be used, so that materials with high mobility have to be selected. The combination of these two conditions more or less limits the choice to the element silicon, which is the commonest, and the binary compounds GaAs and

CdTe for solar photocells. One particular case involves their use as a photovoltaic *detector* of radiations, mainly in the infrared range. In this case, the very slight potential difference at the photocell terminals is applied to a high-impedance amplifier and is thus for practical purposes not delivered. For the close infrared range, from 0.8 to 1 μm, silicon is used; for medium infrared, from 1 to 10 μm, the compounds InAs ($E = 0.4$ eV), PbTe ($E = 0.3$ eV), and InSb ($E = 0.2$ eV); for distant infrared, from 10 to 1000 μm, solid solutions based on HgTe ($E = 0.01$ eV) (Suchet, 1963).

For junction rectifiers, only the elements germanium and silicon have been used industrially, although the use of semiconductor compounds is also possible. The reason for this limitation appears to be connected with the higher cost of monocrystallization of compounds, the risk of variations from stoichiometry, and simply the fact that, since elements were used first and proved satisfactory, any change would have to be justified by better performances to make the necessary investments worth while. And compounds do not appear to offer much better performances. Very brief reference will be made to the main technical problems, which are well known, in constructing junction rectifiers: mechanical attachment of the contacts or removal of heat caused by the Joule effect. Figure 7.5 shows, in diagrammatical form, the way in which they were solved for the earliest types of germanium rectifier. The contacts were made of pure tin and indium, protection against the atmosphere was provided by a two-part Kovar box insulated by a glass ring, and heat was evacuated by the "anodic" current supply braids.

7.3. Multiple junction: amplifier

We shall now move on to the case of three regions in the same crystal, in the order P–N–P, to which a potential difference is applied. Let us assume, for example, that a germanium crystal comprises two regions doped with boron and one doped with phosphorus. The P region connected to the positive voltage is usually called the *emitter*, the P region connected to the negative voltage the *collector*, and the N region kept at an intermediate potential the *base*. Let us first consider the operating principles of the two junctions separately. Figure 7.6*a* illustrates the

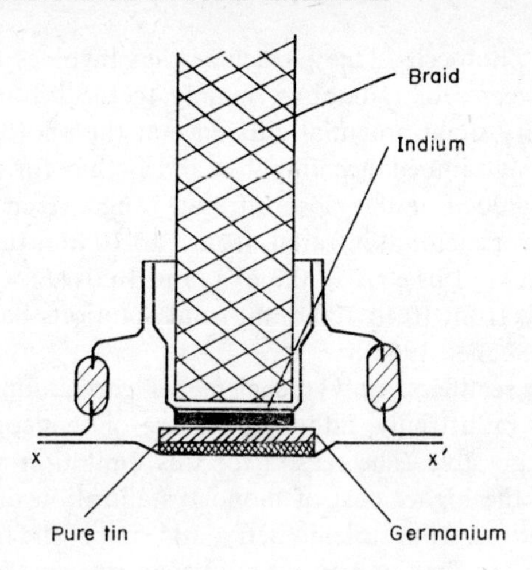

FIG. 7.5. Schema of early types of germanium rectifier.

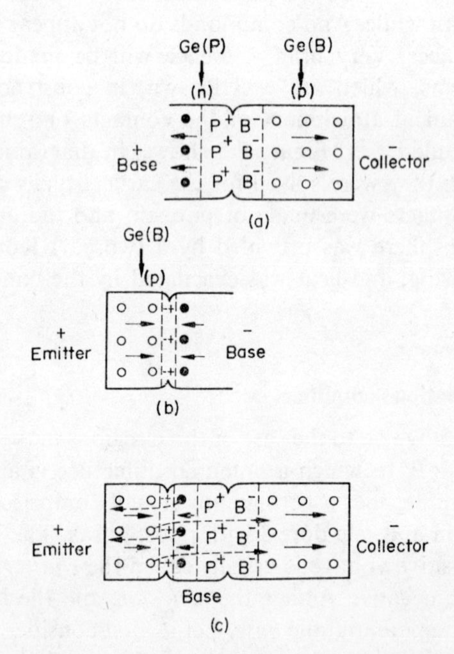

FIG. 7.6. Principle of operation for transistors (based on Suchet, 1962).

main collector–base junction, and the similarity between this diagram and Fig. 7.4*d* immediately shows that it behaves as a rectifier functioning in the reverse direction, and thus letting only a negligible current through. Figure 7.6*b* illustrates the emitter–base junction, and the similarity between this diagram and Fig. 7.4*c* immediately shows that it behaves as a rectifier functioning in the forward direction, and thus allowing a large amount of current through.

Let us now consider interactions between the two junctions (Fig. 7.6*c*). The electrons and holes arriving in the neighbourhood of the emitter–base junction will not all recombine if the distances separating them from the emitter on the one hand and from the collector–base junction on the other are slight, and if the potential difference applied between the emitter and base is high. In this case they will in fact be subject to sufficient electrostatic forces to precipitate a number of electrons through the emitter on to the positive electrode and a number of holes through the base on to the collector–base junction. It is these holes that are essential to the functioning of the device. Speeded up by the local field of the collector–base junction, they reach the negative electrode through the collector, in addition to the small number of holes coming from the collector itself. Very approximately, the conveyance of charges between base and collector can be said to be controlled by the potential difference applied between emitter and base, since it is this difference that controls the behaviour of the emitter–base junction. Since the collector–base impedance is very high (that of a rectifier in the reverse direction), while the emitter–base impedance is very low (that of a rectifier in the forward direction), it can be seen that even if the currents traversing them are approximately the same, the powers involved will be quite different. This is in fact a device for controlling a high output power (collector–base) by means of a small input power (emitter–base), in other words an *amplifier*.

The schematized functioning described above is that of a transistor, which is thus basically a voltage amplifier, whereas the vacuum valves previously used amplified currents. The functioning of the transistor can be seen to be related to the lifetime of the minority carriers, in other words holes in N regions and electrons in P regions, before their recombination. It is mainly the need to obtain high lifetimes that has led to technological advances in the purification of germanium and silicon and their crystallization without dislocations. It will be noted that positive holes play the essential role in the P–N–P transistor taken as an example. In an N–P–N transistor, on the other hand, the functioning of which can easily be deduced from the description above, this role is provided by the electrons, which generally have higher mobilities: hence the usefulness of such transistors at high frequencies.

Transistors made up of semiconductor compounds can function, and have, indeed, functioned experimentally, in accordance with the very brief description above. The mobilities of the minority carriers, however, have to be quite high without E being too low. In practice, this restricts experiments to GaAs, InP, and InAs. SiC, because of the high value of E, could, however, be used in transistors operating at high temperatures, and for this reason has attracted some interest. Finally, amplification is theoretically possible in semiconductors in which the role of the impurities is performed by vacancies or interstitial atoms, but these defects reduce mobility much more than impurities.

Devices using more than three different regions in one crystal are also known, and we shall make some reference here to thyristors, which use four. As for the P–N–P transistor, the positive electrode or anode is connected to the first P region, but the second P region is connected to a control electrode, and acts as a grid. It is a new N region that is connected to the negative electrode or cathode. The thyristor thus consists of two P–N junctions in series, separated by an N–P junction. In the absence of a control current, the device lets no current through in any direction. As soon as a current passes between the grid and cathode, it behaves like an ordinary P–N rectifier and retains this behaviour even if the control current stops. This property allows the rectifier to be triggered off by a single current impulse. Figure 7.7 shows a standard utilization. The *controlled rectifier* RC is inserted between a sinusoidal generator and a charge R. Its control circuit is connected to an impulse generator GI. Depending on the position of the control impulse ig, the rectifier delivers a current I into the charge for the greater part of a half-period, or for a small fraction only. It then cuts the circuit again as soon as its terminal voltage is eliminated, and begins to function again only with the next impulse.

These controlled rectifiers are very suitable for the construction of exciting devices associated with all types of voltage, current, or velocity

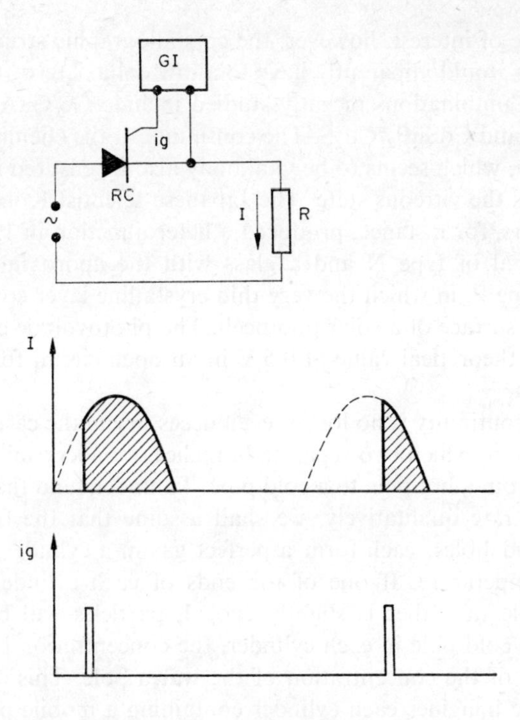

FIG. 7.7. Circuit and operation of a control rectifier.

regulation. There is always a unit responsible for producing a control voltage: an image of the quantity to be regulated, on a suitable scale, is compared with a reference voltage, and the difference is amplified. This voltage controls an impulse generator responsible for producing the control impulses applied to the rectifier grid. One of the commonest applications is the exciting and regulation of the voltage of an alternator.

7.4. Heterojunction: thermocell

The photovoltaic effect and rectifier effect, mentioned in connection with P–N junctions and their applications, also appear in *heterojunctions*, in which the P and N regions belong to different crystals. For the results to be of interest, however, the crystallographic structures of the two crystals would apparently have to allow epitaxic growth of one on the other. Combinations recently studied include Ge/GaAs, Ge/ZnSe, Cu_2S/CdS, and $CdSnP_2$/Cu_2S. The continuity of the chemical bonds at the junction, which seems to be vital, may also be ensured if one of the partners has the vitreous state. The Japanese scientist Komura and his collaborators, for instance, produced a heterojunction in 1970 between a CdS crystal of type N and a glass with the approximate formula As_2S_3 of type P, in which the very thin crystalline layer constitutes the illuminated surface of a solar photocell. The photovoltaic effect almost reaches the theoretical value of 0.5 V in an open circuit for 30,000 lux illumination.

Epitaxic continuity is no longer even necessary in the case of thermoelectric cells, in which two separate branches, monocrystalline or polycrystalline, join a hot pole to a cold pole. To understand their functioning, at any rate qualitatively, we shall assume that the free carriers, electrons and holes, each form a perfect gas in a cylinder at a homogeneous temperature. If one of the ends of each cylinder is slightly heated, while the other is slightly cooled, particles will be displaced towards the cold pole in each cylinder, the concentration increasing at the expense of the concentration of the warm pole. This can be illustrated if one imagines each cylinder containing a mobile piston in the middle of it: such a piston would naturally slide towards the cold pole. If we then connect the ends of each cylinder by a metallic conductor, in which the electric charge carriers circulate at a more or less constant density, the concentrations of the two poles can be allowed to equalize, and the conductor will bring back to the warm pole the same quantity of particles as had been displaced towards the cold pole inside the cylinder.

Figure 7.8 shows two semiconductor units, doped to form types N and P respectively, and consequently similar to the electron and hole cylinders just described. The upper end is warm and the lower end cold. Electric current will accordingly circulate upwards in the left-hand unit, in the opposite direction to the current of negative particles, and downwards in the right-hand unit, in the same direction as the current of positive particles. It is clear that the two units have to be connected in series to deliver into the charge, shown at the bottom of the figure. But the chemical compounds need not be the same in both branches. In fact, the different performances of crystals doped to form N or P types often rule out their use in the same temperature range.

The too main thermoelectric effects are the Seebeck effect, or the appearance of a potential difference under the influence of a temperature gradient, just mentioned, and the inverse effect, known as the Peltier effect, or the appearance of a temperature gradient in a junction of two N and P semiconductors traversed by a current. The first is a source of electrical energy. The second, on the other hand, consumes energy for refrigeration. Theoretically, it could also be used for heating, but it is simpler in this case to use the Joule effect in a metallic conductor.

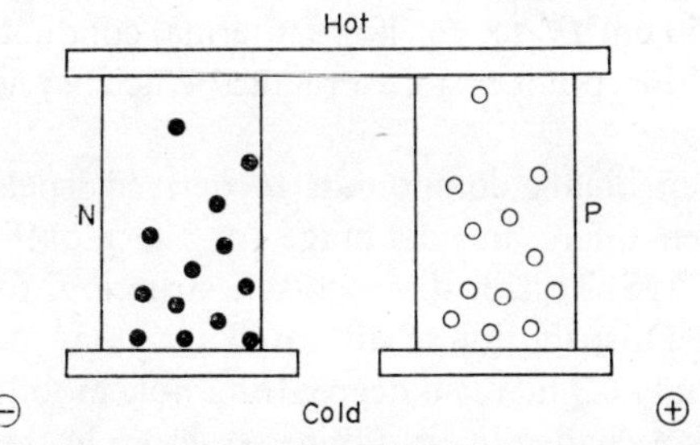

FIG. 7.8. Principle of operation for thermoelectric cells (based on Suchet, 1963).

The two thermoelectric effects are expressed by the equations

$$V = \alpha \Delta T \quad \text{(Seebeck effect),}$$

$$Q = \pi I \quad \text{(Peltier effect),}$$

with $\alpha = \pi/T$. The coefficients α and π are much higher for semiconductors than for conductors, which explains the preference given to semiconductors. In both effects the heat dissipated by the Joule effect is wasted, like the heat transmitted from the warm pole to the cold pole, so that high electrical conductivity and low thermal conductivity are desirable.

Calculations of performances show that it is preferable to use semiconductors with carrier concentrations of approximately 10^{19} per cm^3 and the highest possible mobility/thermal conductivity ratios. In other words, atom arrangements have to be found that will disperse electrons less than phonons. This means the use of a semiconductor material that is still extrinsic at the working temperature, with E ranging from approximately 0.15 eV at around 50°C to 0.3 eV at 300°C and 0.6 eV at 800°C. Sintered materials, where this technique is possible, seem to offer the best performances. In practice, P-type Bi_2Te_3 and N-type Bi_2Te_2Se compounds are used at temperatures close to room temperature and for refrigeration in particular. The mobility of the holes in the first of these compounds is around 500 cm^2/V sec, thermal conductivity is 25 mW/cm degree, and the melting point 586°C. Practical efficiency is approximately 5%. At several hundred Centesimal degrees, and particularly for thermoelectric cells, the P-and-N-type compound PbTe is normally preferred. Mobilities are 1600 cm^2/V sec for electrons and 750 cm^2/V sec for holes, thermal conductivity 22 mW/cm degree, and melting point 904°C. Practical efficiency is approximately 7%.

A change from binary compounds to derived isoelectronic ternary compounds often offers an advantage for the mobility/thermal conductivity ratio. $AgSbTe_2$, in the rock-salt structure, for instance, presents a disordered distribution of silver and antimony atoms. Its thermal conductivity is only 6.3 mW/cm degree for a hole mobility of 360 cm^2/V sec, but it shows a change in crystalline structure in the range in which it is used. The use of solid solutions also allows thermal conductivity to be reduced because of the fluctuation of atomic mass on a given type of crystallographic site. $AgSbTe_2+PbTe$ solutions containing 60% $AgSbTe_2$, for example, have a thermal conductivity of only 4.5 mW/cm degree, and the structure of the ternary is sufficiently stabilized for the structural change to be rejected outside the field of use. Such a solution can accordingly provide the P branch of a device using bismuth-doped PbTe for the N branch. Practical efficiency is approximately 8% between 100° and 500°C under the most suitable conditions (Suchet, 1963).

7.5. Other applications

The Hall effect was explained in § 2.1. The Hall voltage and the current it can result in allow us to regard any four-electrode semiconductor plate placed in a magnetic field as a special device, for which the expression *Hall generator* came into use shortly after the discovery

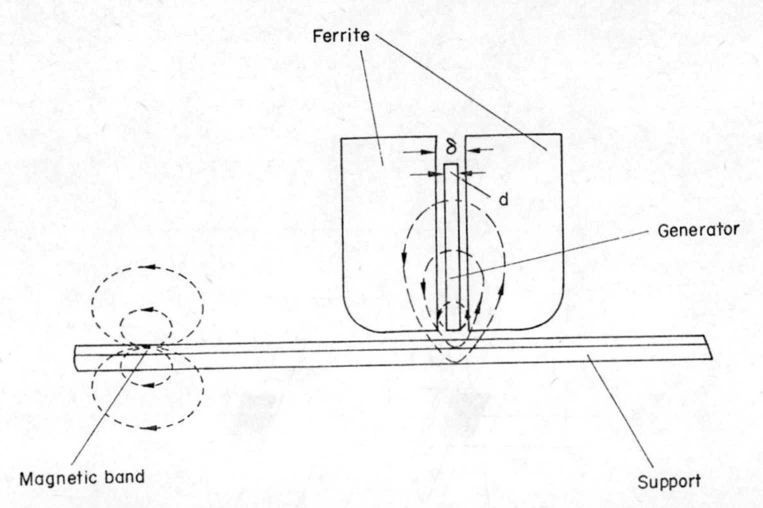

FIG. 7.9. Operation of a magnetic tape-reading head using a Hall generator (based on Weiss, 1964).

of the compounds InSb and InAs, and their exceptionally high mobility, by Welker in 1955.

Hall generators are mainly used to measure magnetic fields, whether the earth's magnetic field, concentrated in a μ-metal rod, the field resulting from the proximity of magnetic objects, the position of which can thus be detected, magnetically coated letters or figures, sound or visual signals recorded on magnetic tapes, or the intensity of electric currents, by means of the magnetic fields they create around them.

Figure 7.9, for example, illustrates the functioning of a magnetic tape-reading head using a Hall generator. Longitudinal magnetization of the tape causes variations in the magnetic flux in a ferrite head made from a magnetic material not possessing any remanent magnetic moment, but with high permeability, usually mixed Mn–Zn or Ni–Zn ferrites. Heads exists in which the effective gap δ does not exceed 10 μm and the thickness d of the Hall generator, produced by cathodic spraying in a vacuum, a few microns.

Figure 7.10 shows the principle of the measurement of a high current I by means of Hall generators powered by an auxiliary current I_0. The Hall voltage V_H at the terminals of the two generators, assembled in

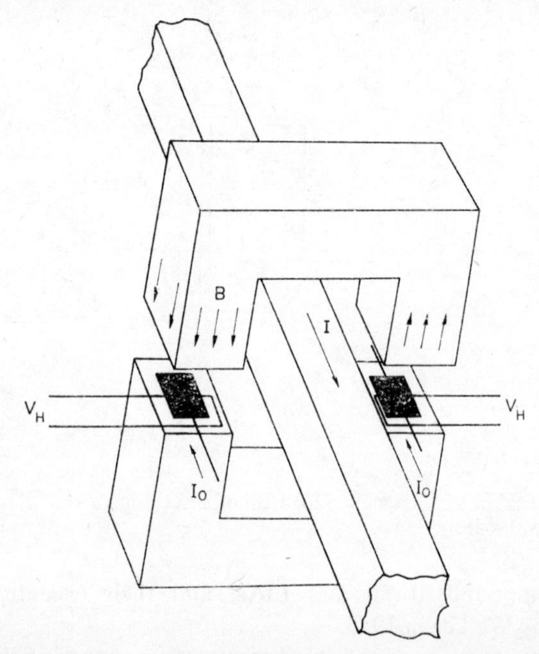

FIG. 7.10. Principle of the measurement of high intensities using Hall generators fed by an auxiliary current I_0 (based on Weiss, 1964).

series, is proportional to the magnetic induction B in the two-piece magnetic core surrounding the electrical conductor. Such cores are usually made of soft iron in contrast to the heads described above, since they are not traversed by high-frequency current. The dimensions of the circuit and width of gaps are calculated so that the soft iron in the circuit is saturated, corresponding to a field of approximately 10,000 Oersteds in the gap. The advantage of saturation is the slight influence of magnetic fields induced by other electricity lines or nearby ferrous objects on measurement. Accuracy is around 0.2%.

Another application is based on the tunnel effect of certain P–N junctions. This paradoxical effect consists of the passage of a current in a junction such as that in Fig. 7.4c, while the positive potential applied to the P region is very low and the corresponding field is still lower than the local field (cf. § 7.2). Under these conditions, carriers in the two regions wishing to pass through the junction come up against a potential barrier which their pure corpuscular aspect would not allow them to surmount. The intervention of the wave associated with any material corpuscle, which follows the laws of wave mechanics, allows this apparent anomaly to be explained in the same way as the laws of reflection and refraction in geometrical optics have to be reinterpreted in physical optics to explain, for example, interference phenomena. The effect is possible if the semiconductor is strongly doped on each side of a very thin junction.

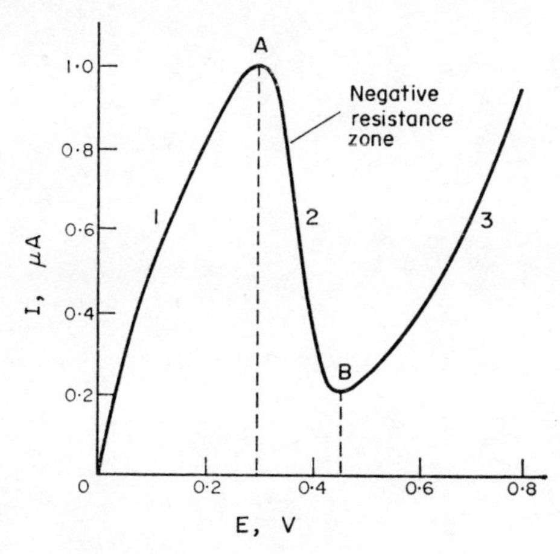

FIG. 7.11. Current–voltage characteristic in the forward direction of an Esaki "tunnel" diode.

Figure 7.11 shows the *I–V* characteristic discovered by Esaki for certain types of semiconductor rectifier. Zone 1 involves the tunnel effect because, in view of the value of the potential applied, current should still be nil. Zone 2 corresponds to the end of the tunnel effect and the return to normal conditions. Zone 3 corresponds to the *I–V* characteristic of the rectifier operating normally in the forward direction. Zone 2 accordingly corresponds to functioning in which the junction has a negative resistance, a quality appropriate for the provision of an *oscillating circuit.* The velocity of passage of electrons through the barrier in this zone is fairly close to that of light, so that frequencies of around 10,000 MHz can be used.

References

SUCHET, J. P. (1962) *Chimie physique des semiconducteurs*, Dunod, Paris (English translation, van Nostrand, London, 1965).

SUCHET, J. P. (1963) *Curso de compostos químicos semicondutores* (*Course on Chemical Semiconducting Compounds*), Centro Técnico de Aeronautica, São José dos Campos, SP, Brazil.

WEISS, H. (1964) *Solid-State Electron.* **7**, 279.

Chapter 8

Magnetic Semiconductors

8.1. Thermistors

The variation in relation to temperature of the resistivity of metals (§ 1.2) and of semiconductors (§ 2.1) has already been described, and reference was made in § 6.4 to the use of this variation in nickel or platinum thermometers. If the temperature coefficient of the resistance is defined as $\alpha = d\rho/\rho \, dT$, the characteristic relations of metals and semiconductors show that $\alpha = 0.4\%$ per degree approximately for the former, while $\alpha = -E/2kT^2$, namely -1 to -5% per degree at room temperature for the latter, according to the value of E. It is therefore clear that it is preferable to use semiconductor materials. However, these can operate only within a reduced temperature range, since the coefficient drops very quickly when the temperature T increases. Such components are called "negative temperature-coefficient resistors" (NTC) or *thermistors.*

Allusion to this application in the chapter devoted to magnetic semiconductors may seem surprising, since it is in no way connected with the existence of a magnetic order. There is obviously no theoretical reason not to use conventional semiconductor materials to make thermistors. Their low resistivity and high cost, however, usually form technical and economic obstacles to their use. In practice, the only materials on the market are, on the one hand, manganites $NiMn_2O_4$ and $Ni_{0.5}Co_{0.5}Mn_2O_4$, which behave like the non-stoichiometric oxide NiO and CoO, and, on the other hand, haematite Fe_2O_3 with valency induction by TiO_2. In both cases the semiconduction mechanism is based on transfers of electrons on a d sub-level, between Ni^{II} and Ni^{III} or between Fe^{II} and Fe^{III}, and the carrier mobility is very low. The same applies to the few other materials also proposed. In addition, preparation of these elements in the form of sintered aggregates allows one to benefit from a fortuitous circumstance, connected with the particularities of electrical conduction in granular structures with a resistant or conducting surface layer. It can be shown that, for ceramics of oxides NiO or CoO produced by sintering in air, the presence on the surface of a more conducting layer (resulting from a high density of metal vacancies) results in a variation in resistivity in relation to temperature, of the form

$$\rho = AT^b \exp(B/T),$$

where b is negative: hence a temperature coefficient of the resistance

$$a = b/T - B/T^2.$$

It can be seen that the drop in the coefficient, as the temperature rises, occurs slightly less rapidly than would be required by the normal law for semiconductors. For example, a coefficient of -5% per degree at 25°C is still -2.9% per degree at 150°C if $b = -5$.

The main applications of thermistors are, as might be expected, in thermometry and thermoregulation. Their working temperature ranges are usually confined to 200° or 250°C: higher temperatures are preferably measured by means of thermoelectric couples. A "bridge assembly" is commonly used to compare the thermistor with an ordinary resistor, the output diagonal comprising a deviation reading instrument or electronic recording or regulation appliance. The current passing through the thermistor is limited by its thermal dissipation, and must cause only a negligible temperature rise so as not to affect the accuracy of measurement. A typical example is the measurement of the temperature of water in motorcar radiators. Competing devices are the thermocontact with bimetallic strip, which gives only an alarm signal, and the vapour-pressure thermometer, connected by a tube to a small pressure gauge on the dashboard, inconvenient in the case of rear-engined cars. Figure 8.1 shows how a pellet can be placed in oil inside a standard measuring device, with the inside spring forming the contact. An element of 500 Ω at 25°C and 35 Ω at 100°C dissipating 0.5 W without significant temperature rise is suitable. The measurement appliance is a quotient-meter with two windings, one connected directly to the battery terminals so as to eliminate the effect of variations in the voltage supplied by it.

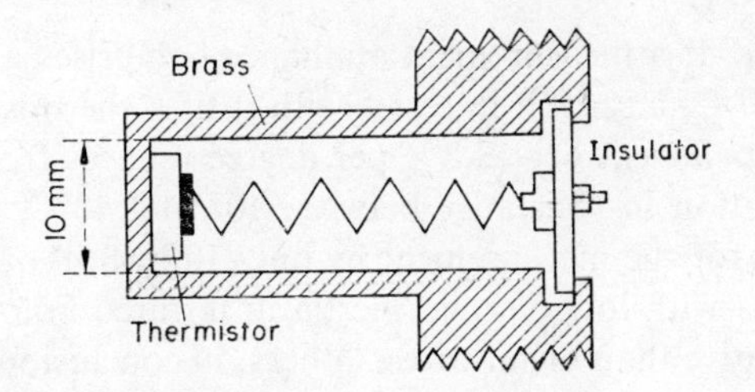

FIG. 8.1. Mounting of a thermistor inside a standard device to measure the temperature of an internal combustion engine radiator (based on Suchet, 1955b).

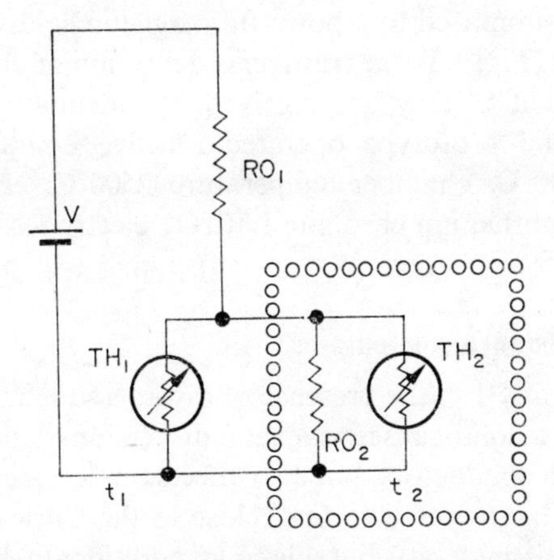

FIG. 8.2. Diagram of the regulation of a small heat-insulated chamber at temperature t_2 for room temperature t_1 (based on Suchet, 1955b). RO = ordinary resistor, TH = thermistor.

Heat regulation of small enclosures can also make use of very simple circuits, such as the one shown in Fig. 8.2, comprising ordinary resistors RO and thermistors TH. The apparatus includes a wire resistor RO_1, a heating winding RO_2 and two high-dissipation thermistors TH_1 and TH_2. If the characteristics of these components and the voltage source are suitably selected, the temperature t_2 in the heat-insulated enclosure should remain stable for a variation in the room temperature t_1 or the arrival of calorific energy in the enclosure at temperature t_2.

Other applications involve compensation of the low positive coefficient of metals. If a measurement appliance comprises a copper frame of 100 Ω at 20°C, insertion in series with it of a thermistor of 10 Ω at 20°C with a coefficient of −3.8% per degree reduces from 12 to 2.5% the total variation in resistance between 10° and 40°C, increasing the power needed for the measurement by only 10%. Still others make use of components with low thermal inertia in infrared bolometers, which are more sensitive than metal wires. Others, in conclusion, make use of the time needed for heating of the thermistor to delay the action of a relay.

It is possible to add here refractory semiconducting materials used as electrodes in magnetohydrodynamic electric generators. A high-temperature gas, ionized by caesium, is sent from a supersonic tewel into a chamber submitted to a powerful magnetic field. As in the case of the Hall effect (Fig. 2.3), the transverse deflection of electrons creates an electric current if they are received by adequate electrodes. A Russian industrial prototype operated for five hours in 1974: gas temperature 2000°C, chamber temperature 2500°C, chamber section 110 × 40 cm, lanthanum chromite $LaCrO_3$ electrodes (f^0_a, see § 5.1), efficiency 50–60%.

8.2. Magnetoresistance commutator

As was stated in § 1.4, the presence of disordered magnetic moments on the atoms of a solid causes magnetic dispersion. This phenomenon is not confined to conductors, but also affects semiconductors, in which one should also be prepared to find, close to the Curie or Néel point, an increase in resistivity, attributable to an additional term of magnetic origin. Since, in contrast to metallic conductors, the number of N or P carriers can vary (if the conventional semiconduction mechanism intervenes), it is not surprising to see it appearing in this term:

$$\rho_{\text{spin}} = KS(S+1)/N.$$

As for the principal term of resistivity, it has already been seen in § 3.2 (Fig. 3.5) that a special semiconduction mechanism, by transfers of d electrons from one atom to another, usually predominated on the conventional mechanism and determined its value. This mechanism is similar to the one referred to in § 7.1 (Fig. 7.3), which may appear at low temperatures in the impure crystals of conventional semiconductors

$$\rho_{\text{principal}} = \rho_0 \exp(E_S/kT).$$

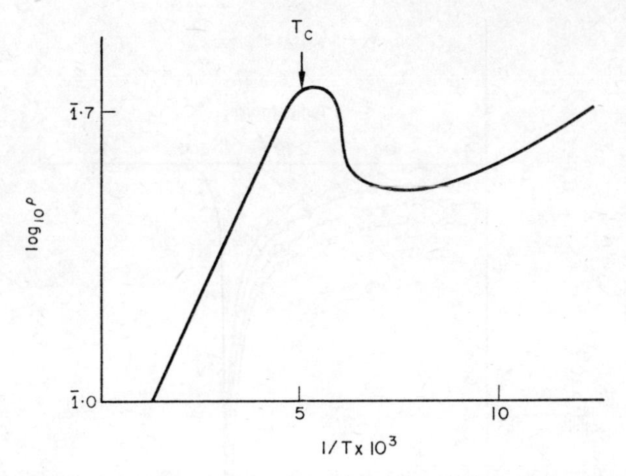

FIG. 8.3. Variation in the logarithm of the resistivity of $FeCr_2S_4$ in relation to the reciprocal of absolute temperature (based on Goldstein and Gibart, 1969). T_C = Curie point.

At low temperatures, far away from the disorder of atomic moments, exponential decrease in the principal term is undisturbed. Immediately below the Curie point, the sharp increase in the magnetic term attenuates the decrease in the principal term, and may even compensate for it, if the transfer energy E_S is low. At higher temperatures, once disorder of the moments is established, exponential decrease is resumed.

Such curves were in fact found experimentally between 1961 and 1965 by various authors, with the ferromagnetic dioxide CrO_2, of C4 structure. Its monocrystallization is very difficult and it decomposes around 250°C, namely only 130° above its Curie point. Research into its conduction mechanism is accordingly extremely difficult (Suchet, 1971). It was seen in § 3.3 that even if, as it seems, it is metallic, it constitutes an exception to Goodenough's rule and an extreme case, since small additions are enough to cause semiconductor properties to appear. It may be, therefore, that insufficient purity in the samples caused the appearance of a very low transfer energy. One curve probably less open to doubt was published in 1969 by Goldstein and Gibart (Fig. 8.3). It concerns the ferrimagnetic thiospinel $FeCr_2S_4$, the transfer energy E_S of which is also very low, around 0.02 eV. If applications are to be found for these properties, the portion of the curve corresponding to a

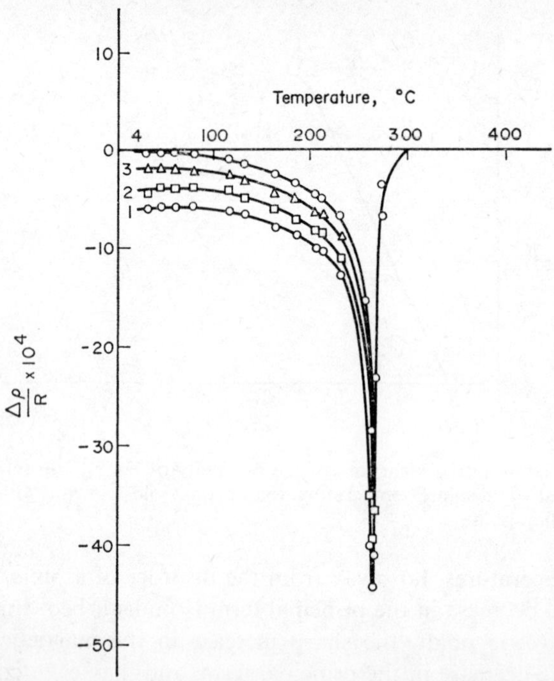

FIG. 8.4. Variation in the magnetoresistance of $MnFe_2O_4$ in relation to temperature in saturating fields of (1) 1960, (2) 1700, (3) 1439, and (4) 1178 Oersteds (based on Belov and Talalaeva, 1957).

high positive temperature coefficient of resistivity would have to be large enough. The type of application alluded to in § 6.4 (baretter) occurs as a possibility. The succession of portions of curve with positive and negative coefficients has also been considered (Suchet, 1955a).

The concept of magnetoresistance was introduced in § 1.3 in connection with metallic conductors. The general lines of its variations in relation to temperature or the applied field are the same in semiconductors. For example, Fig. 8.4 shows, for different fields, the effect of temperature on the ferrite $MnFe_2O_4$ in which the algebraic minimum is particularly sharp. But whereas $\Delta\rho/\rho H$ at the Curie point scarcely exceeds 0.005 in metals and alloys, a value of 0.02 was pointed out in 1949 for CrTe, and values of several tenths have recently been reported by Holtzberg *et al.* (1964) for the compound EuSe, and by Haas *et al.*

(1967) for the selenospinel $CdCr_2Se_4$. So under what conditions has this effect, for which the expression "giant magnetoresistance" has been created, obtained? It is now known that it can occur only in a semiconductor, but that it requires a high transfer density in a narrow band. This was obtained in EuSe by partly replacing europium (f^7) by gadolinium (f^8), in which valency II is not stable. This causes, in each gadolinium atom, transition of an electron to the 5d level, where there is soon a high transfer density. In selenospinel, the divalent cadmium in the *A* sites was partly replaced by trivalent gallium. This resulted in valency induction on the B sites, where divalent chromium ($d\epsilon^3 \alpha d\gamma^1 \alpha$) appears alongside the trivalent chromium ($d\epsilon^3 \alpha$), and the level $d\gamma^1 \alpha$ is partly filled. In both cases, the Ruderman–Kittel–Yosida interaction also allows the Curie point to be raised: 45°K for EuSe (cf. § 3.3) and 122°K for selenospinel.

Magnetoresistance devices at present exist in the electronics industry, but they use diamagnetic materials such as InSb, bismuth (cf. § 6.4), or other metals, in which the effect is very slight. The materials mentioned above are much more promising, notably $Cd_{0.98}Ga_{0.02}Cr_2Se_4$, in which the magnetic field needed would be less than 1 kOe. If the Curie point could be increased to around room temperature, and the peak, still too sharp, flattened out, the way would be open for *commutators* without mechanical contacts, which would offer a high level of reliability.

Commutation by magnetoresistance could be used to convert a.c. into d.c. and for low-frequency modulation and oscillations. However, we believe that the most interesting applications lie in the improvements made possible in low-powered d.c. motors. Such motors offer very valuable advantages, such as a high starting torque and wide range of operating velocities. In their conventional form at least, however, they also involve two major drawbacks: the fragility of the components providing revolving contacts, and the variation in speed under load, in the absence of regulation. The appearance since the Second World War of permanent magnets of oxides with a high coercive field suggests their possible use as a rotating inductor or rotor, as is already common in light motor-cycle dynamos. But in this case the different windings on the fixed armature of the motor, namely the stator, need to be commutated in relation to the position of the magnet, namely the value of the field. Giant magnetoresistance components could perform this function, whether by being inserted directly into the windings, or by controlling silicon thyristors or transistors.

8.3. Special magnetoelectric effects

There are two such effects: the extraordinary Hall effect and the Astrov effect. What they share is that, to our knowledge, no application has so far been envisaged. But no one can forecast the future, and it seems advisable to say a few words about them.

The Hall effect has already been mentioned in connection with conventional semiconductors, §§ 2.1 and 7.5. This was the ordinary Hall effect, and a description of it was fairly straightforward. We now have to consider the more general case of a semiconductor plate in which a magnetic order prevails. The Hall voltage V_H to be measured between the opposite side surfaces of the plate (cf. Fig. 2.3) comprises two terms, the first of which is of the same nature as the Hall effect in diamagnetic materials, and depends on the magnetic induction B created in the plate by the field H to which it is subject. It is generally this term that has the smaller value. The second term is new, and depends on the magnetization intensity M in the plate. Let us assume that the plate is square, with 1 cm sides, and of thickness a. We have

$$V_H = [R_0 B + R_1 M](I/a).$$

R_0 is the ordinary Hall coefficient, previously referred to simply as R; it was seen that the mobility μ of the majority carriers was merely the product $R\sigma$ or $R_0\sigma$. R_1 is the extraordinary Hall coefficient, about which much less is known.

The two terms can be separated if the plate can be saturated easily, which implies that its thickness a is not too small. Above saturation, the second term remains constant and the slope of the quantity aV_H/I in relation to B gives the coefficient R_0. Figure 8.5 shows the different cases that can occur, depending on the respective signs of the coefficients R_0 and R_1. The curve represented in this figure is the initial magnetization curve. There is naturally a hysteresis cycle if the applied field disappears and changes its sign.

Theoretically, the same applications are possible as for the ordinary Hall effect, although the hysteresis is a drawback; but since very little

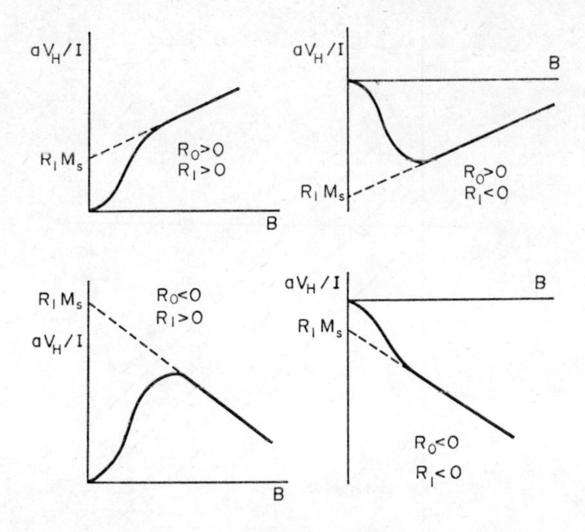

FIG. 8.5. Various possible forms of the variation in aV_H/I in relation to induction $\boldsymbol{B}$ (based on Suchet, 1971).

is still known about the extraordinary effect, it is not absolutely impossible that a new application may some day appear. We therefore tried, a few years ago, to find a criterion enabling performances of various magnetic materials to be compared. Without going into calculations in detail, it might simply be mentioned that, if the voltage V that has to be applied to it to make the current I pass is made to appear, we obtain

$$V_H/V \sim \mu\mu^* H.$$

The ratio of the voltage received to the voltage applied is thus proportional to the magnetic field applied and to a factor of merit $\mu\mu^*$, the product of the mobility of the majority carriers by an effective magnetic permeability related to the coefficients R_0 and R_1. Figure 8.6 is a double logarithmic graph in μ and μ^*, which shows the points or figurative regions of various materials: conventional semiconductors InSb and HgTe, low-magnetic solid solutions such as GeTe/MnTe, metals, and ferrites.

The Astrov effect is the induction of a magnetic moment in a crystal

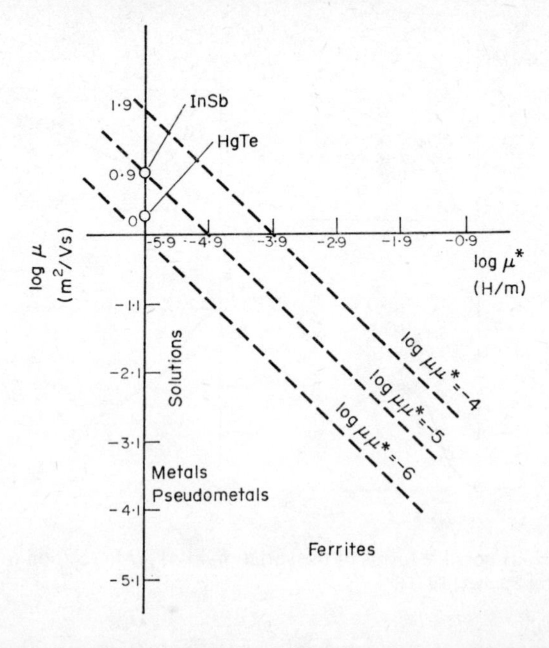

FIG. 8.6. Double logarithmic graph: mobility of majority carriers–effective magnetic permeability (based on Suchet, 1966).

by an electric field. It was predicted on a purely theoretical basis by Landau and Lifshitz, and linked to the existence in thermodynamic potential of an invariant expression

$$\mathbf{B} = \mu\mathbf{H} + \alpha\mathbf{E}$$

where **H** and **E** are the magnetic and electric field vectors. It was discovered in 1959 in the antiferromagnetic oxide Cr_2O_3 by Astrov, who found an interaction coefficient α of 6×10^{-4} in the direction parallel to the ternary axis, which remains very slight. The inverse effect, the induction of an electric moment by a magnetic field, is also observable, and is linked to the expression

$$\mathbf{D} = \epsilon\mathbf{E} + \alpha\mathbf{H}.$$

Both effects arise from the physical phenomenon of distortion of the electron cloud round an atom carrying a magnetic moment.

Rado has recently observed an interaction coefficient ten times higher in the compound $Ga_{2-x}Fe_xO_3$ close to $x = 1$, which shows slight ferromagnetism. It may therefore be hoped that its observation in a strongly ferromagnetic material, if possible, will give high enough interaction coefficients for the electro-inductive effects derived from it to have practical applications. These include in particular the construction of inductance coils or capacities variable by remote control, by the application of polarization or magnetization.

8.4. Magneto-optical effects

An isotropic transparent medium can become anisotropic under an outside influence such as pressure or an electric or magnetic field. Two separate indices then appear for a given direction of light propagation, the ordinary index n_0 and the extraordinary index n_e. This is the phenomenon of *birefringence*. Any incident ray therefore gives two rays in the medium involved, the ordinary ray and the extraordinary ray, which propagate at different velocities, so that after a distance l the corresponding vibrations show a path difference

$$\delta = (n_e - n_0)l$$

and a phase difference

$$2\varphi = 2\pi\delta/\lambda,$$

where λ is the wavelength of the monochromatic light used.

Magnetic birefringence was discovered at the end of the nineteenth century by Cotton and Mouton, in nitrobenzene,

$$n_e - n_0 = C\lambda B^2,$$

where B is the magnetic induction of the medium being investigated and C, the Cotton–Mouton constant, is approximately 10^{-12} e.m.u. in nitrobenzene. A more important effect has recently been observed in the paramagnetic crystal EuF_2, but ferromagnetic transparent media can give an intense effect, e.g. garnet $Y_3Fe_5O_{12}$ at room temperature or EuSe at the temperature of liquid helium (Suits and Argyle, 1965).

An anisotropic transparent medium, placed in a magnetic field, also causes a rotation in the polarization plane of a light ray passing through it in the direction of the field. This magnetic *rotatory power*, discovered by Faraday in 1846, has been explained by Fresnel as a difference in absorption between the right-hand and left-hand circular vibrations, corresponding to the indices n and n'. This is what is called circular dichroism. After a distance l, this phenomenon results in a path difference

$$\delta = (n'-n)l$$

and a phase difference 2φ, causing the polarization plane to rotate by an angle

$$\varphi = C'lB,$$

where C' is the Verdet's constant. Rotation changes direction at the same time as the field, so that a ray passing forwards and backwards through the medium considered, after reflection off the other surface, shows a phase difference 2φ. This represents a major difference compared with natural rotatory polarization in which, under the same conditions, there would be no phase difference, since the special molecular structure responsible for the phenomenon cannot change direction. The constant C' is approximately 0.01–0.1 min/cm G in common liquids, but attains −9.7 min/cm G for 0.67 μm in the paramagnetic crystal EuSe, and enormous values in the garnet $Y_3Fe_5O_{12}$.

There is a strong resemblance between the optical absorption curves and those for specific magnetic rotation, and in particular the two quantities increase rapidly up to an absorption threshold corresponding to the wavelength $1.24/E$ (eV), expressed in microns. This was observed very clearly in the garnet $Y_3Fe_5O_{12}$, and more recently in the three chromium trihalides at low temperatures. The same d or f electrons would therefore appear to be responsible for both magnetism and optical absorption. We have already seen, in connection with rare-earth compounds, the important part played by f–d transitions. The magnetic rotatory polarization should therefore be used in the immediate neighbourhood of the threshold. Anyway, this is where maximum transparency is to be found. Figure 8.7 shows the absorption threshold on the left, and on the right the beginning of a rise in absorption according to λ^2, fairly slow at (*a*) in the case of a pure crystal and faster at (*b*) if the crystal is less pure.

Magnetic rotation of the polarization plane is also observed in Kerr's magneto-optical effect, which concerns the reflection of light on the polished surface of a magnetized part such as the pole of an electro-magnet. It is known that reflection on an insulating surface such as glass introduces a phase difference π. On a metal surface, reflection introduces a complex phase difference, the value of which differs for the two inverse circular vibrations, and this effect is sensitive to magnetization of the surface. Rotation amounts to approximately 20 min for saturated iron. A much more intensive effect was recently observed

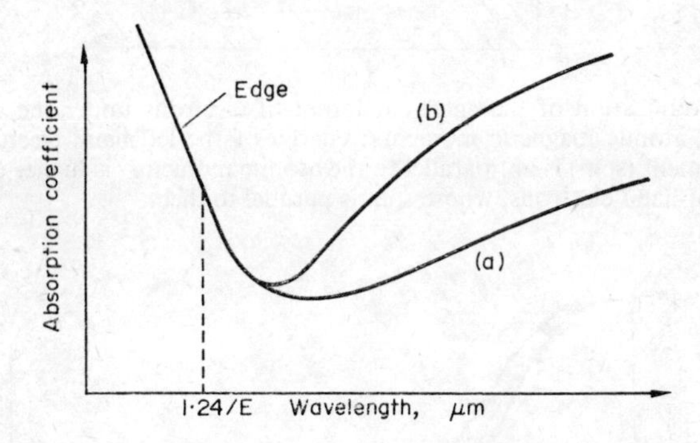

FIG. 8.7. Variation in the optical absorption coefficient in relation to wavelength in a pure (*a*) or impure (*b*) semiconductor crystal (based on Suchet, 1971).

by Greiner and Fan (1966) on polished EuO and EuS crystals at low temperatures, in other words in the ferromagnetic range. This effect is connected with the 4f–5d electron transitions in the europium atom.

Finally, a quite new magneto-optical effect, theoretically predicted for several years, was observed experimentally in 1964 by Busch *et al.* in EuO at a low temperature. Figure 8.8 shows, in diagrammatical form, how the magnetic order partially splits the bonding system of energy levels (occupied by valency electrons) and the antibonding system (occupied by free electrons) in relation to respective orientations of electron spin moments and atomic moments. The interval E between the two systems is then decreased by an amount ϵ which depends on the intensity of magnetization in the crystal. The effect is usually

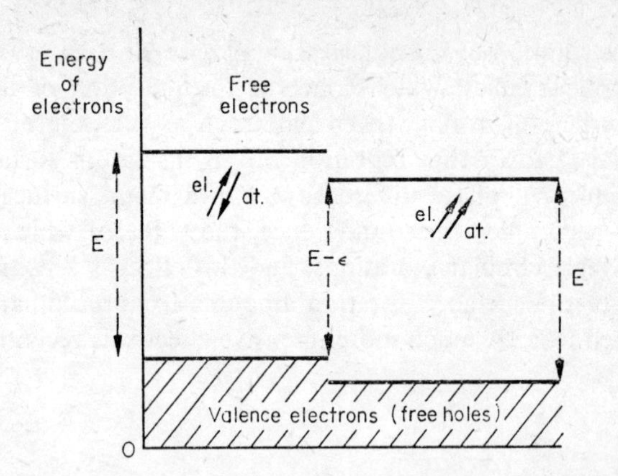

FIG. 8.8. Modification of the energy diagram of electrons under the influence of ordered atomic magnetic moments: energy of the left-hand electrons, whose self-moment (spin) is antiparallel to the atomic moments, is higher than that of the right-hand electrons, whose spin is parallel to them.

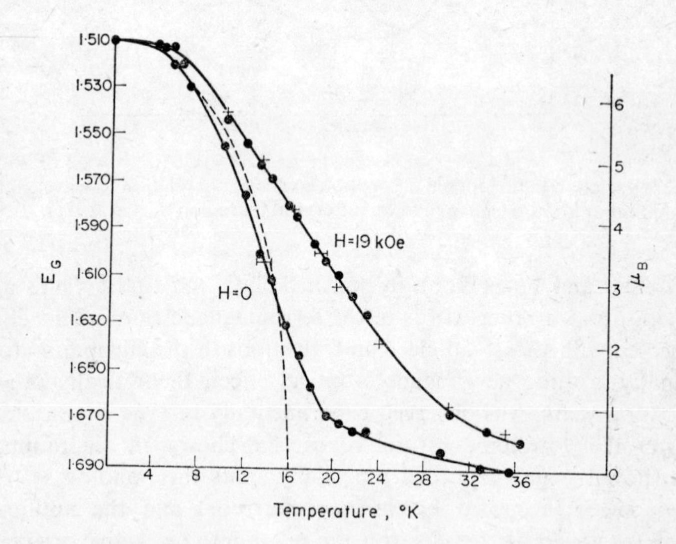

FIG. 8.9. Variation in E, deduced from the optical absorption threshold, and the average number of magnetons per europium atom in relation to temperature (based on Busch and Wachter, 1966). ● = experimental results; + and - - - - = Brillouin curves $B_{7/2}$, $\theta = 16°K$ for 11.2 and 0 kOe.

two- to three-tenths of an electron volt, and corresponds to a *threshold shift* towards red, in accordance with earlier predictions. It has also been observed in EuS, EuSe, and other crystals. Figure 8.9 shows the phenomenon in the case of EuS, and also gives the value of the average atomic moment at each temperature, with or without an external field. The same team has recently observed a similar displacement in the antiferromagnetic compound EuTe, amounting to three-hundredths of an electron volt, but this time towards blue. It would thus be theoretically possible, by modifying the value of the applied magnetic field, to vary the transparency of a magnetic semiconductor for a suitably selected wavelength.

8.5. Laser modulator

The Faraday effect, or magnetic rotatory polarization, can be used to deflect or modulate electromagnetic waves. Certain one-directional ferrites or garnet guides, developed in the fifties, had already used it to deflect hyperfrequency waves. Recent use of gas lasers, which at present form the most powerful continuous source of coherent light, in atmospheric windows of 3–4 μm (helium–neon) and 9–10 μm (carbon dioxide), and the wish to modulate their rays to transmit data have caused fresh interest in electro-optical and magneto-optical effects. Of magneto-optical effects, the Faraday effect seems to be the most suitable to use for modulation.

Several modulator systems have been proposed, and Fig. 8.10 illustrates the one patented in 1957 by Dillon. The sample is placed in a waveguide in the X band (namely approximately 3 cm) and led by klystron to the ferromagnetic resonance frequency. A static magnetic field H saturating the sample is applied perpendicular to the path of the light ray of frequency F, so that there is no polarization-plane rotation for any frequency. On the other hand, at the resonance frequency f, a precessional motion round the direction of H is imparted to the saturation magnetization vector $\mathbf{M}_S$, and the polarization plane rotates in proportion to the component m_S. After passing through the analyser, an amplitude modulation of the frequency F to frequency f is obtained. If the frequency f of the klystron has previously been frequency-modulated by the information to be transmitted, there is no obstacle to its remote reception. A plate of garnet $Y_3Fe_5O_{12}$, for example, can be used with a resonance frequency of 10–100 GHz. Tests have also been carried out with $CrBr_3$ and EuSe. In general, development of this process is less advanced than modulation by Pockels' electro-optical effect, but it offers different characteristics.

We have seen that the relation existing among activation energy, absorption threshold, specific rotation maximum, and transparency maximum requires selection of a crystal in which the wavelength of the absorption threshold is immediately below that of the light for modulation. This leads to a gain in transparency and in the Verdet constant.

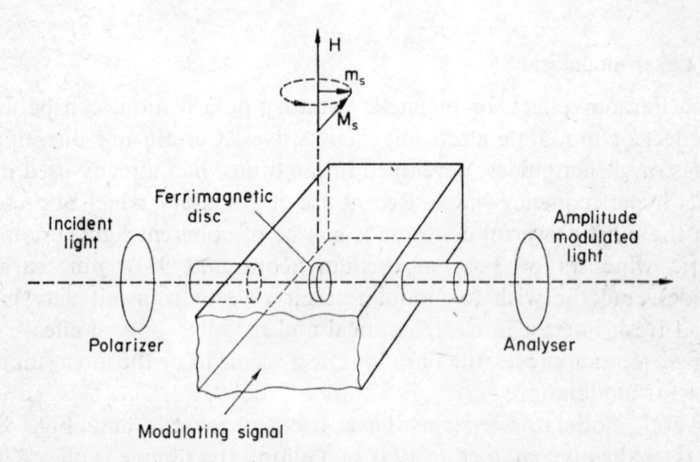

FIG. 8.10. Diagram of a Faraday-effect modulator operating at the ferro- or ferri-magnetic resonance frequency (based on Dillon, 1957).

Accordingly, E should be approximately 0.5 eV for the helium–neon laser emitting at 3.39 μm, and 0.15–0.2 eV for the carbon-dioxide laser emitting at 9–10 μm. Of magnetic semiconductor materials at present known, $Y_3Fe_5O_{12}$ is more appropriate for 0.85 μm and $CrBr_3$ for 0.48 μm. Only thio- and selenospinels have lower activation energies, and their use has recently been suggested for 10 μm. The MCr_2S_4–MCr_2Se_4 series, where M is cadmium or mercury, and the TCr_2S_4 series, where T is manganese, iron, or cobalt, are now known.

Dillon's patent concerned only amplitude modulation, and this system presents drawbacks. Infrared laser communication systems are mainly intended for use in zones where condensation and smoke frequently obscure the atmosphere, and could therefore be partly absorbed, even at the selected wavelengths. A recent patent modifies the original system, to provide for frequency modulation. To do this, it sends the laser wave F, amplitude-modulated by the hyperfrequency wave f, into a second laser which acts as an amplifier, but is set to $F \pm f$ or $F \pm 2f$. To understand this, one should remember that a wave of frequency F, amplitude-modulated to frequencies f or $2f$, can be broken down into a Fourier series, so that alongside the fundamental frequency F are found the two satellite frequencies $F \pm f$ or $F \pm 2f$. It is one of them that is selectively amplified for the second laser, with the result that amplitude modulation is changed into frequency modulation.

Figure 8.11 illustrates these modulation problems more clearly. At the top, the fundamental frequency F of 30,000 GHz, or *carrier* frequency, constitutes the coherent radiation of the carbon-dioxide laser. Below, the auxiliary wave f, approximately 60 GHz, or *sub-carrier*, represents the wave of a hyperfrequency generator, frequency-modulated in the klystron by the information to be conveyed. The frequency of this sub-carrier must be selected to correspond exactly to the difference existing between two successive emission lines of the carbon-dioxide laser. Further down is the rotation of the polarization plane, which is obviously proportional to the preceding frequency. Below this again is the signal observed at the amplifier laser input. Two cases are possible, depending on the regulation of the plane of extinction of the analyser in relation to the polarization plane of the beam, in the absence of modulation: amplitude modulation can occur at frequency f or at frequency $2f$. Finally, at the bottom of the figure, the waves in both cases are shown broken down into Fourier series. The shape of the wave at the amplifier laser output is not shown. Its amplitude would be constant, but its frequency would vary slightly, around $F+f$, for example, with modulation frequency resulting from the information to be transmitted. Even if it is a television picture, this frequency is only around 100 MHz, and accordingly cannot be represented alongside the value of 30,060 GHz of $(F+f)$.

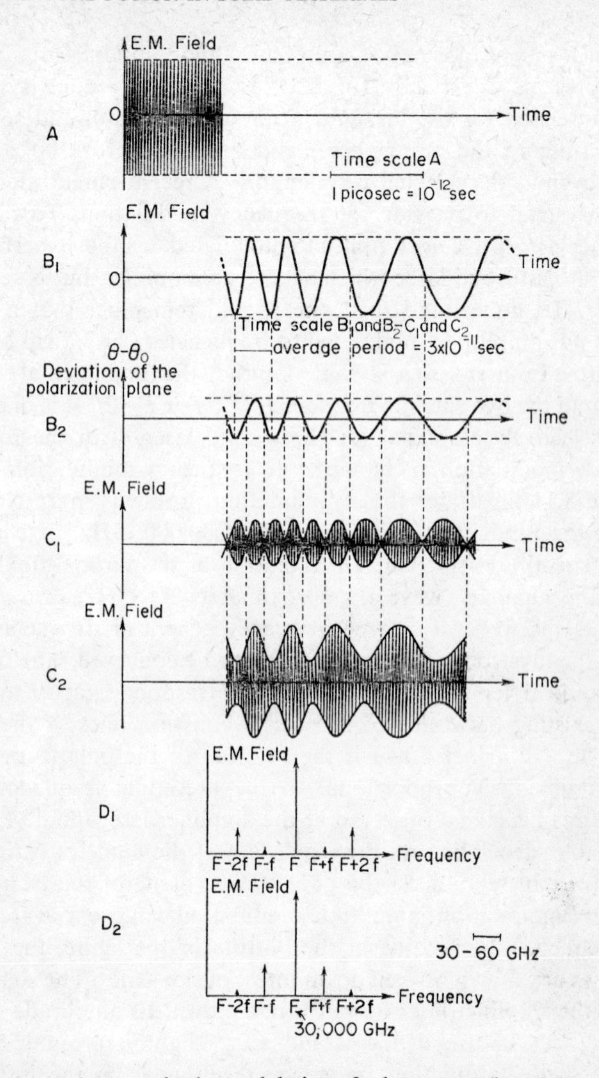

FIG. 8.11. Successive steps in the modulation of a laser wave (based on Suchet and Sage, 1970). A = laser wave (carrier); B_1B_2 = high-frequency wave, modulated by the information (sub-carrier); C_1C_2 = amplitude-modulated laser wave (two cases possible); and D_1D_2 their breakdown into Fourier series.

References

ASTROV, D. N. (1960) *Zh. eksp. teor. Fiz.* **38,** 984.
BELOV, K. P. and TALALAEVA, E. V. (1957) *Zh. eksp. teor. Fiz.* **33,** 1517.
BUSCH, G., JUNOD, P. and WACHTER, P. (1964) *Phys. Lett.* **12,** 11.
BUSCH, G. and WACHTER, P. (1966) *Phys. Kondens. Mater.* **5,** 232.
DILLON, J. F. (1957) American patent 2,974,568 (15 February 1957).
GOLDSTEIN, L. and GIBART, P. (1969) *C.R. Acad. Sci. Paris* **269 B,** 471.
GREINER, J. H. and FAN, G. J. (1966) *Appl. Phys. Lett.* **9,** 27.
HAAS, C., VAN RUN, A. M., BONGERS, P. F. and ALBERS, W. (1967) *Solid State Commun.* **5,** 657.
HOLTZBERG, F., MCGUIRE, T. R., METHFESSEL, S. and SUITS, J. C. (1964) *Phys. Rev. Lett.* **13,** 18.
RADO, G. T. (1964) *Phys. Rev. Lett.* **13,** 335.
SUCHET, J. P. (1955a) *Bull. Soc. fr. Electns. Paris* **5,** 274.
SUCHET, J. P. (1955b) *Les Varistances et leur emploi dans l'électronique moderne,* Chiron, Paris.
SUCHET, J. P. (1966) *C.R. Acad. Sci. Paris* **262 B,** 127.
SUCHET, J. P. and SAGE, M. (1970) French patent 2,076,313 (9 January 1970).
SUCHET, J. P. (1971) *Crystal Chemistry and Semiconduction in Transition Metal Binary Compounds*, Academic Press, New York.
SUITS, J. C. and ARGYLE, B. E. (1965) *Phys. Rev. Lett.* **14,** 687.

Chapter 9

Switching Semiconductors

9.1. Thermal detector

Devices to measure temperatures and detect variations in them have always made use of the physical phenomena most sensitive to such factors: the emission spectrum for high temperatures, linear expansion, then electrical conduction of metals, and, finally, of semiconductors (cf. § 8.1) for medium temperatures. But there are cases in which an even faster variation is needed, no longer for an extensive temperature range, but within an interval of only a few degrees. Bimetallic strips (cf. § 6.5) offer an excellent solution to this problem in the mechanical field, but an electrical solution may be preferred, notably to ensure greater reliability. For this reason, magnetic or crystallographic transitions are beginning to arouse interest.

In the semiconductor field it is crystallographic transitions that cause the sharpest variations in resistivity. It was seen, in § 4.2, that this was the case, for example, for magnetite at 120°K. This oxide could perfectly well have been used to set off a signal when the transition temperature was reached, or even to regulate the temperature of an enclosure close to 120°K, if this temperature had been of particular interest. This is not the case, however, and transition temperatures situated close to room temperature or within a range of 100° approximately above room temperature are needed. The semi-conductor–metal transition of V_2O_3 at 170°K involves the same drawback, namely that of lying outside the normal temperature range.

The transition that occurs in VO_2 at 68°C, on the other hand, corresponds to the conditions laid down, and it is undoubtedly at the basis of the operation of the "critesistors" or critical-temperature

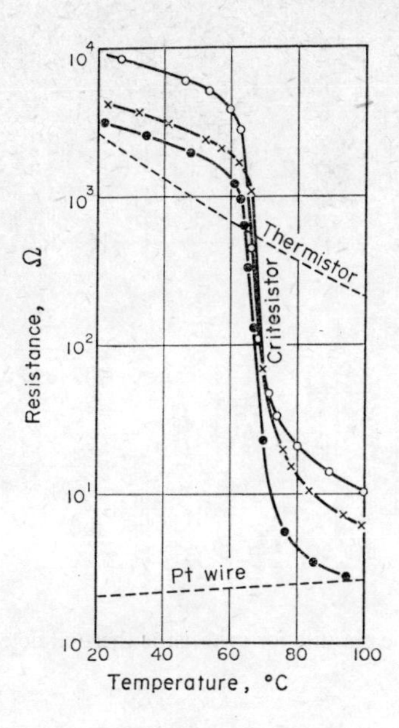

FIG. 9.1. Compared resistance–temperature characteristics of a platinum wire, a thermistor, and some critesistors (based on Futaki, 1965).

resistors studied by Futaki and marketed in Japan. This involves a mixture of V_2O_5 and stable-valency oxides sintered in a reducing atmosphere, and which shows a very sharp variation in resistivity. The effect of stable-valency oxides is probably a valency induction bringing part of the vanadium back from valency V to valency IV. In the transition zone of these components, the negative temperature coefficient of resistivity is thirty times higher than in normal thermistors, as Fig. 9.1 shows. Such a characteristic arouses fresh interest in conventional applications of thermistors, and in particular those using them as nonlinear components with low calorific capacity in the form of beads deposited on two fine platinum wires.

In the neighbourhood of the transition temperature, critesistors have great sensitivity to temperature. It can be shown that the electric input power required by a sensitive component is inversely proportional to the *square* of the temperature coefficient. Figure 9.2 shows two diagrams of temperature-regulating circuits particularly suitable for the constant-

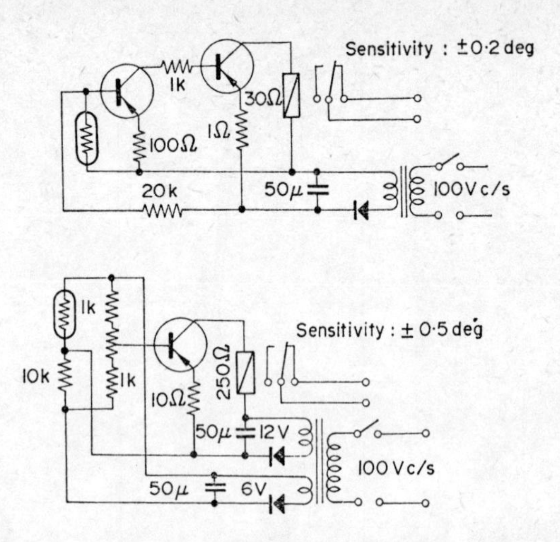

FIG. 9.2. Possible electric circuits for critesistor thermal detectors (based on Hitachi Ltd., 1965).

temperature chambers used for semiconductor crystals in measurement and communication equipment.

Another application of the transition of VO_2 has appeared quite recently. This involves direct display of time on watch dials. The surface of the dial is divided into numerous parts, each comprising a small monocrystalline plate heated by a minute electrical resistor. The plates are cut along a plane designed to ensure that variations in the optical index on each side of the transition will cause the greatest difference in reflecting power. Various combinations of these plates accordingly allow digital display of the time for an almost negligible expenditure of energy.

9.2. Memory switch

These components are useful in applications derived directly from the phenomenon of reversible crystallization described in § 4.4. They are constructed by instantaneous evaporation of crystallizable glasses in a vacuum and depositing of a film approximately 1 μm thick on two small graphite hemispheres, which are then placed in contact over an

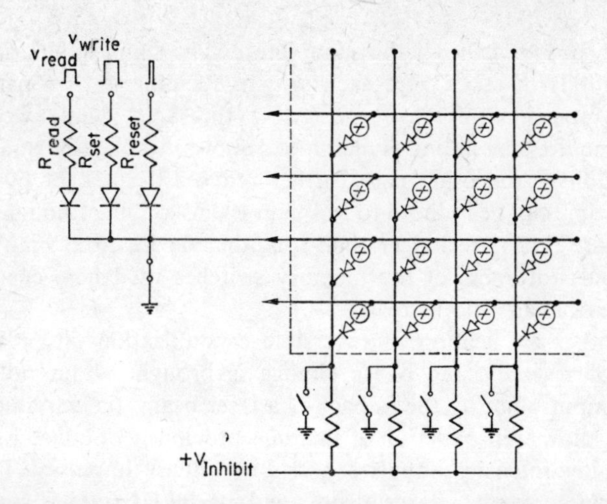

FIG. 9.3. Diagram of a network of diode-insulated chalcogenide glass memory switches (based on Nelson, 1970). $V_{\text{read}} < V_{\text{threshold}}$, $I_{\text{read}} \sim 0.5 I_{\text{set}}$, $V_{\text{inhibit}} = V_{\text{write}} > V_{\text{threshold}}$.

area of approximately 0.3 mm^2, both films being face to face. Their initial resistance, which is greater than 200 kΩ, can drop to around 100 Ω ("set" state), after a voltage impulse which is above the threshold (current of 2–20 mA for a few milliseconds), and return to the initial state ("reset" state) after a current impulse of 100–400 mA for approximately 10 μsec, in other words short enough for localized cooling to take place quickly.

Memory switches can easily be used in a network of perpendicular conductors, where they provide a connection between horizontal and vertical lines at each crossing point (Fig. 9.3). In such a device, the applied voltage between a horizontal and vertical line does not cause a current to pass solely through the switch located at their intersection: the current can also take a roundabout path, using another horizontal and another vertical. It then has to pass through a series of three switches, but the more extensive the network the larger will be the number of these parallel bypasses. It can be seen that this drawback has to be avoided to ensure that there is no ambiguity in the change in position of the switch, in other words the *writing* of a figure in binary language, or in order to be able to detect such a change subsequently in other words *reading* of the same figure. The simplest solution is to place a diode in series with each switch. Reading is then performed with a voltage impulse lower than its threshold, while writing or erasure requires the impulses mentioned above. Application of a reverse polarization to the other cells further reduces the risk of bypassing. Reading time can be reduced to less than 100 nsec and, although writing and erasure times are longer than for other devices, the simplicity of design and sturdiness of the memory switches used here ensure that such a device can hold its own.

In another application of reversible crystallization, at present the subject of research, the phase change is brought about not by an electric impulse but by the impact of a laser beam, the variable power of which allows reproduction of shadings previously obtained by means of long, low-intensity impulses or short, intense impulses. The thin film of glass covers a large drum, and its initial state is uniformly microcrystalline. It is scanned by the beam of a laser operating continuously and electronically modulated by a computer or by television signals. For example, the resistivity of a film of selenium–tellurium solid solution rises from 10^6 to 10^{12} Ω cm at points melted by a 1.5 W laser and cooled quickly in the vitreous phase. Points of only 2 μm in diameter have thus been recorded. The image is stable, consisting of resistant points or zones, and it can be transferred to paper as many times as required by standard electrostatic duplicating processes. Heating to a moderate temperature recrystallizes the whole film.

Figure 9.4 shows a diagram for the practical performance of this new printing process. The black and white image stored on the layer of glass is perfectly stable in the absence of heating by the laser. Once a non-conducting latent image has been formed, the surface is placed in contact with a pigmented powder which adheres electrostatically to the non-conducting regions. The powder is then transferred to paper and the process can be repeated, still with the same image.

In view of the very recent nature of research into reversible crystallization, development of its applications should be regarded with caution. Memory switches exist on the market, however, and operate satisfactorily. In fact, licences have been granted and commercial agreements drawn up concerning them. Application to electronic memories has been developed with the help of the American Army, which was

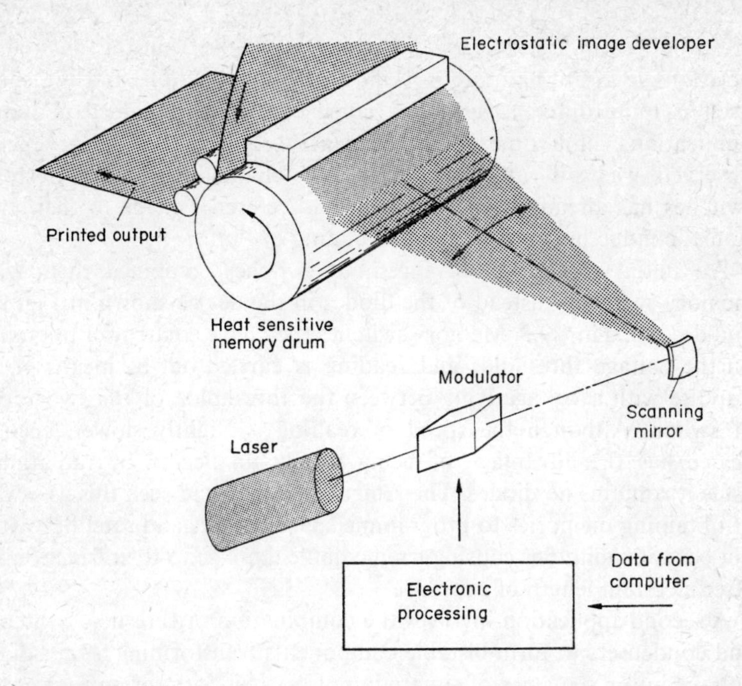

FIG. 9.4. Diagram of an electrostatic printing system in which a laser beam modifies, by pinpoints, the state of a semiconducting amorphous film (based on Simmons, 1970).

interested in one property of these glasses not yet mentioned: their insensitivity to cosmic radiation. Such radiations are a considerable obstacle to the use of transistors in space. Here, too, such memories exist, and have been presented in various purely commercial circles. The printing process, in contrast, is still being developed. It has functioned in a prototype, but is not yet produced industrially. It could probably be industrialized only by a firm specializing in duplicating processes.

9.3. Threshold switch

These components are used in applications derived directly from the phenomenon of non-destructive breakdown alluded to in § 4.5. They are produced in the same way as memory switches, but the glass compositions used are naturally specific to this application, and the effect itself is quite different, although some confusion remained in initial publications, at a time when the exact origin of non-linear effects observed was still unknown. The *I–V* characteristic of threshold switches has already been discussed, and reference made to their two states: conducting or "on", and insulating or "off".

An initial application of threshold switches combined them with memory switches, instead of the diodes in the device shown in Fig. 9.3 and described in § 9.3. Memory switches then have a narrower tolerance on the voltage threshold, and reading is carried out by means of an impulse with a voltage lying between the thresholds of the two types of switch. Although the speed of reading is slightly slower, such a device has the advantage of being totally unaffected by radiations, since it contains no diodes. The American Army had seen this as a way of obtaining memories to programme space rockets and satellites without cosmic radiations causing irremediable damage to their functioning after a certain length of exposure.

A second application involves the combination of threshold switches and condensers to form bistable components transforming a sinusoidal voltage into a sequence of square impulses. This form of power supply allows to be kept uniformly brilliant an electroluminescent lamp, which is a particular type of condenser. In addition, a charge or discharge of the condenser can shift the voltage range within which it operates, causing ignition or permanent extinction of the lamp (Fig. 9.5). Let us assume that the switch 1 is momentarily closed. The condenser discharges immediately. Subsequently, it will follow the threshold switch without difficulty, polarizing in one direction or the other with each square impulse. The light is thus lit. Let us now assume that switch 2 is momentarily shut. The condenser takes in its maximum charge. It can then no longer react to alternations with the same sign. Only alternations with the opposing sign will reduce its charge slightly, but not enough to produce brilliancy in the lamp, which is thus extinguished.

Many attempts have been made in the past to arrange a large number of electroluminescent cells and their circuits on a flat panel, notably in order to retransmit television pictures, but these have always come up against the problem of fading of the contrast. The brilliancy of a given

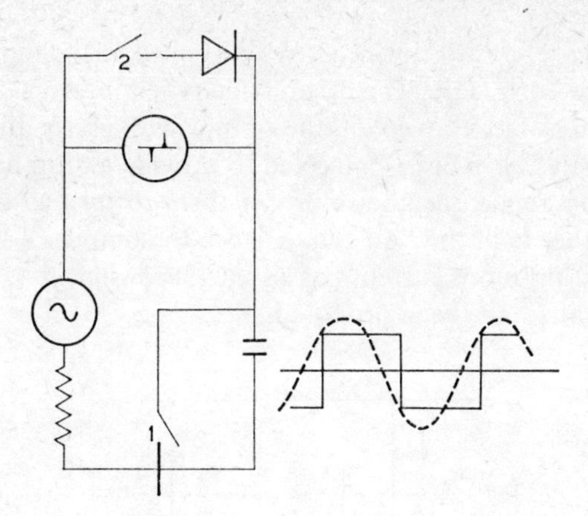

FIG. 9.5. Feeding of an electroluminescent lamp comprising a threshold switch (based on Fleming, 1970).

cell varies only rather slowly in relation to the applied voltage, so that widely differing voltages have to be applied if there is to be a real change from extinction to ignition. With the type of circuit just described, this problem disappears. In addition, if the images are to remain unchanged for some time, which is the case with remote setting-up, brilliancy has to be kept intact, and this is more easily attained with square signals, which contain a high proportion of high-frequency sinusoidal components, more active in the functioning of the cells.

This second application is still at the laboratory stage, for one very simple reason: electroluminescent cells require high supply voltages, whereas the threshold voltages of thin-film switches are low. This problem seems unlikely to remain unsolved for very long, so that we feel it worth mentioning the application here.

Finally, other potential applications of threshold switches exist in certain logic electronic circuits. This type of application usually requires a component with three connections in which a power gain is obtained between input and output. The use of two threshold switches in series has been suggested, input being located at the shared point and output between the device and the charge resistor. If the threshold of the switch is at three-quarters of the voltage applied to the whole system, they share this voltage equally and both remain "off". If a voltage impulse, positive or negative, is applied to the input, one switch moves to the "on" state. The other, which then receives the whole voltage, does the same, connecting the output with earth. In Fig. 9.6, two of these devices have been connected to provide a "flip-flop" assembly. The coupling condenser allows one of them to turn to the "off" state when the other is in the "on" state, since the output of the first one is earthed, and no more current passes into its switches. Other examples of logic circuits have been proposed.

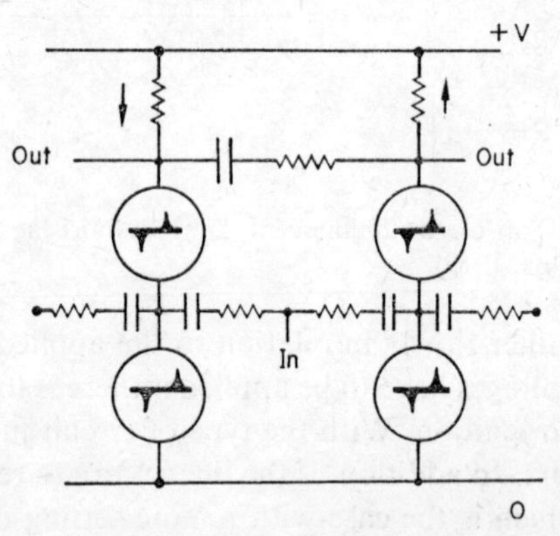

FIG. 9.6. Flip-flop mounting using two threshold switches (based on Nelson, 1970).

All these applications should be regarded as mainly speculative, since the corresponding components are not yet available on the market.

9.4. Special glass IRdome

We have now completed consideration of applications and possible applications of semiconductor–metal transitions. However, chalcogenide glasses have also been discussed, and it would be regrettable if their other applications were not mentioned, the main one being based on their wide transmission in the infrared range. The extent of the transparency of a solid to electromagnetic waves is limited on one side by the absorption threshold and on the other by wavelengths capable of causing the crystal lattice or elementary groups of atoms, in the case of a glass, to vibrate. For an ionic crystal the basic absorption resulting from grouped exciting of the atoms frequently occurs in the close infrared range. In glasses, on the other hand, the position of the first absorption bands depends on the most electronegative element, and moves quickly away towards high wavelengths when oxygen is replaced by sulphur, selenium, and tellurium. Oxide glasses rarely transmit beyond 3 μm. For chalcogenide glasses, the position of the first band is given below in relation to the nature of the chalcogen and its tri- and tetravalent neighbours. In practice, transmission windows of 0.6–11.5 (sulphides), 1–15 (selenides), and 2–20 μm (tellurides) are possible for suitably selected compositions.

	P	As	Si	Ge
S	11 μm	14.2 μm	9.5 μm	12.8 μm
Se	14.3	21.2	15.5	19.3
Te	20	33	20	33

The need for transparent materials in infrared optics was modest for a long time. The glass As_2S_3 has been known since 1870, and its preparation by the reaction of the elements followed by distillation in nitrogen was found quite satisfactory. It transmits well up to 8 μm and its softening point is 200°C. The selenide As_2Se_3 was also known. The need for more elaborate materials appeared about 1960, when certain military devices, such as the homing heads for anti-aircraft missiles, were being developed. The "hot points" formed by the outlet apertures of aircraft jets are pursued by means of the radiations they emit in the 3–4 μm window, and increasingly in the 8–14 μm window. In addition, the components protecting infrared detectors, known as infrared domes or *IRdomes*, can now encounter temperatures of above 200°C because of air friction at supersonic speeds.

Research carried out—for other reasons—from 1955 on in the Soviet Union and from 1959 on in the United States was accordingly encouraged by military experts, and glasses such as Ge–As–Te and Si–As–Te, which have softening points of 450°C, were studied. Figure 9.7 shows the principle of the infrared autodirectors that from 1960 equipped the earth-to-air missile *T*8 (USSR) and the air-to-air missiles Falcon and Sidewinder (USA), Firestreak (UK) and *M*-100*A* (USSR). In this figure, *A* represents the IRdome, which was not always made of glass; M_1M_2 the catadioptric system; *B* a mobile diaphragm; *D* a detector consisting of four cells, the asymmetrical lighting of which modifies the trajectory of the missile; *P.A.* a preamplifier; *E* the electronic circuits. In the civil field it will be possible to replace the rock-salt prisms in spectrometers and build interferential filters.

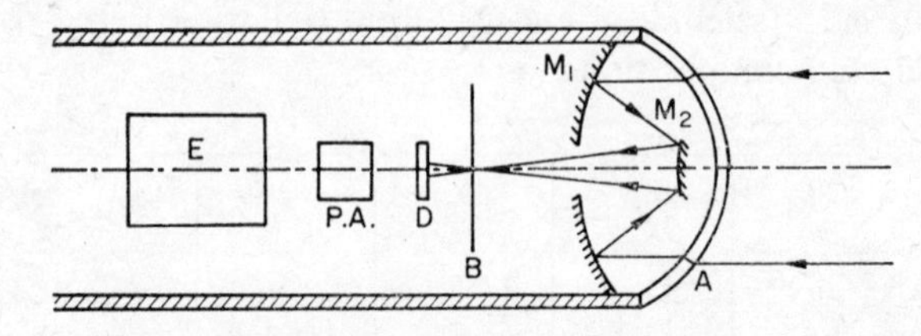

FIG. 9.7. Diagrammatic drawing of an infrared autodirector (based on Hadrot 1960).

The main difficulty involved in preparing these glasses for optical applications is the total elimination of oxygen by distilling the elements in hydrogen. Figure 9.8 shows the difference in transmission spectra depending on whether or not such distillations have been carried out for a $Ge_{34}As_8Se_{58}$ glass under a thickness of 1.8 mm (at *a*) and for a $Ge_{10}As_{50}Te_{40}$ glass under a thickness of 1.62 mm (at *b*).

9.5. Other applications of glasses

Of other possible applications suggested for chalcogenide glasses, we shall mention only those that have begun to be implemented: infrared detectors and acousto-optical devices.

Photoelectric effects are widely used to detect light, notably photoconductivity, or the creation of two carriers of opposing signs by a grain of light. The phenomenon increases with the applied electric field until the mean free path of the carriers thus created is greater than their distance from the electrodes. It has mainly been studied in Se, Cu_2O, Tl_2S, Sb_2S_3, and, more recently, in PbS, PbSe, InSb, and doped

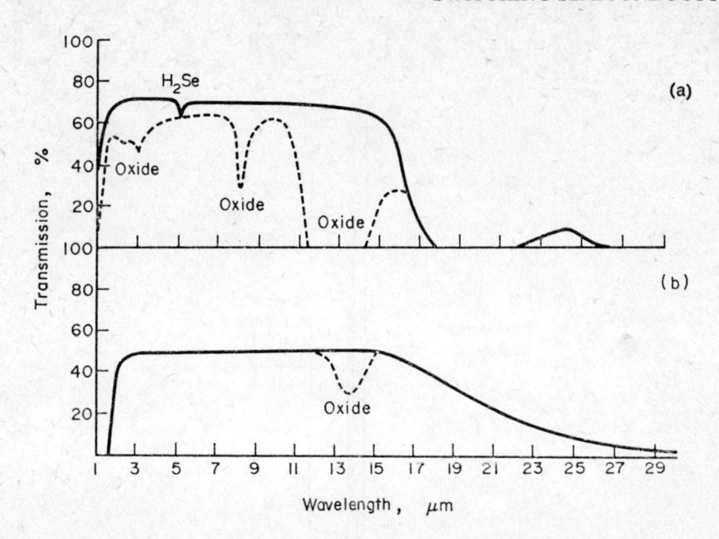

FIG. 9.8. Optical transmission of distillation-purified glasses: (*a*) $Ge_{34}As_8Se_{58}$, (*b*) $Ge_{10}As_{50}Te_{40}$. The dotted-line curves refer to impure samples (based on Savage and Nielsen, 1965).

germanium. Figure 9.9 shows a front view of a Vidicon receiver tube, which uses a thin photoconductor film, on to one side of which the image to be analysed is projected, through a transparent electrode at +20 V, and which is scanned on the other by an electronic beam propagating inside a vacuum tube. The scanning charges the film to the negative potential of the emitting cathode, and neutralization of this charge through the photoconducting substance, at points that are lit, produces a signal proportional to illumination.

The material of the thin photoconductor film must have a resistivity of around 10^{12} Ω cm, so that the relaxation time is greater than the interval of 1/30th sec separating two successive scannings. Good results have been obtained with Sb_2S_3, and an improvement observed when the film was deposited in an amorphous form, which occurs, for example, in a poor vacuum. In 1959 Kolomiets and Lyubin found that the sensitivity of vitreous films of As_2Se_3 was still greater than that of Sb_2S_3 films. In view of the known resistance of glasses to radiations, Vidicon tubes equipped with As_2Se_3 films were thereupon selected to retransmit pictures televised from the satellites Vostok 2 and Soyouz 3,

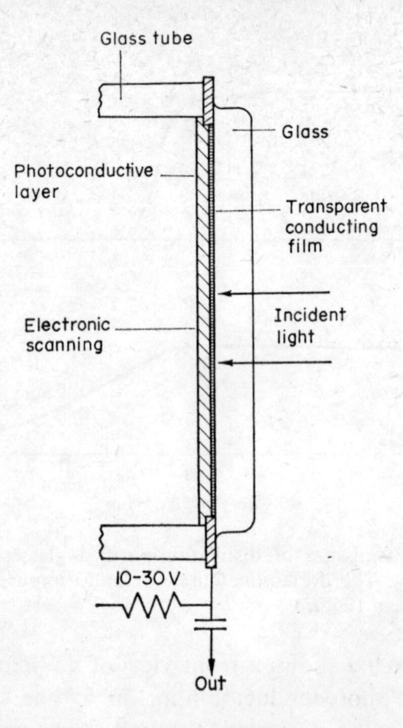

FIG. 9.9. Principle of operation of a photoconductor-film Vidicon camera (based on Weimer *et al.*, 1950, 1951).

and from the moon vehicle Lunakhod. Special glasses have also allowed X-ray-sensitive Vidicon tubes to be produced, and these are being developed in the Soviet Union in the metallurgical industry.

The development of acousto-optical devices is much more recent: it was in 1970 that Krause *et al.* pointed out the special properties of certain chalcogenide glasses such as Ge–As–S: very low acoustic losses, low sound-velocity values, and very high acousto-optical factors of merit. Such properties could be used in ultrasonic delay lines and infrared acousto-optical devices.

Only three atoms per hundred of arsenic need to be added to the binary glass GeS_2 for the longitudinal attenuation of sound at 20 MHz to drop well below 1 db/cm (decibel per cm), and the longitudinal attenuation for the glass $Ge_{30}As_5S_{65}$ is around 0.2 to 0.3 db/cm be-

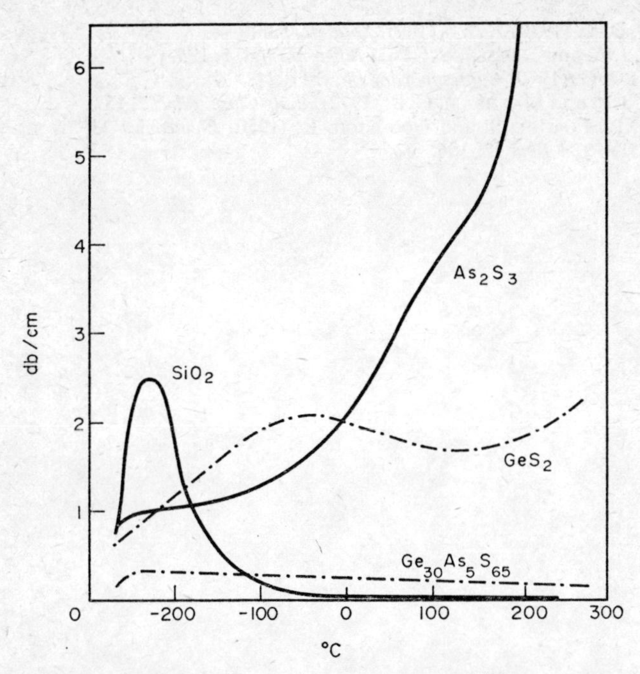

FIG. 9.10. 20-MHz longitudinal acoustic attenuation in relation to temperature for some chalcogenide and oxide glasses (based on Krause *et al.*, 1970).

tween −250° and +300°C (Fig. 9.10). At room temperature it increases with frequency much as in molten silica. The speed of sound in it is around 2.5×10^5 cm/sec, density around 4 g/cm^3 and the optical index high, around 3, so that the acousto-optical factor of merit in these glasses makes them appropriate to modulate and deflect light from infrared lasers (cf. Chapter 10).

References

FLEMING, G. R. (1970) *J. Non-Cryst. Solids* **2,** 540.
FUTAKI, H. (1965) *Jap. J. Appl. Phys.* **4,** 28.
HADROT, M. (1960) *Docaéro* (*Fr.*) no. 60, 23.
HITACHI LTD. (1965) Publicity sheet CS-E061 (January 1965).
KOLOMIETS, B. T. and LYUBIN, V. M. (1959) *Fizika tverd. Tela* **1,** 740, 899.
KRAUSE, J. T., KURKJIAN, C. R., PINNOW, D. A. and SIGETY, E. A. (1970) *Appl. Phys. Lett.* **17,** 367.

NELSON, D. L. (1970) *J. Non-Cryst. Solids* **2,** 528.
SAVAGE, J. A. and NIELSEN, S. (1965) *Infrared Phys.* **5,** 195.
SIMMONS, J. G. (1970) *Contemporary Phys.* **11,** 21.
SUCHET, J. P. and MAGHRABI, C. E. (1972) *Annls Chim. Paris* **7,** 157.
WEIMER, P., FORGUE, S. and GOODRICH, R. (1950) *Electronics* **23,** 70, May issue; (1951) *RCA Rev.* **12,** 306.

Chapter 10

Insulators

10.1 Electrotechnical insulator

The electricity lines described in § 6.1 require high-voltage insulators at the points at which they are mechanically carried by pylons. Although glass has appeared, the commonest material is porcelain. In § 5.1 it was explained that its resistance is very high, and surface enamelling increases impermeability and resistance to corrosion. Its compressive strength is 35–50 kg/mm^2 and its tensile strength 3–5 kg/mm^2. Mechanical and electrical breakdowns are not independent of each other, and the dielectric strength is always lower when measured under mechanical strain. The shape of insulators is designed to ensure that any arc that forms will preferably flash out towards the outside of the insulator, but also so that leaks along the surface will follow the longest possible path, which is partly sheltered from rain. The electrodes must be sealed in position with sufficient plasticity to allow for cements expanding in damp weather.

Up to 10,000 V rigid insulators are made from a single piece, weighing several hundred grams. Above this level, several parts, sealed together with cement, are fired separately, allowing more homogeneous treatment than for a single piece weighing more than 1 kg. In addition, the low electrical conduction of cement makes it play the role of an equipotential surface, ensuring better distribution of the potential drop. In practice, above 30,000 V, preference is given to a chain of suspension insulators, of the "cap and rod" type, as shown in Fig. 10.1. Even if there is a defect in one insulator, sufficient insulation is still left, and this is important for safety purposes. Breakdown voltages are more or less proportional to the lengths of the chains thus assembled, and equivalent, under dry conditions, to those for the same air interval. Rain places the chain at a disadvantage, because of the increased surface conductivity of the insulators.

In small electrical switchgear (plugs, switches, fuses, various lighting and household items), porcelain has now been almost entirely replaced by plastics, since voltages are usually 220 V, and no excessive temperature rise is to be feared. The commonest resins are phenolic resins, obtained by condensing a phenol with an aldehyde: phenol–formol, cresol–formol, phenol–furfurol, etc. Aminoplasts, obtained by urea–formol, thiourea–formol, and melamine–formol condensations, are also used, as well as cold mouldings, by compression, of asphalt-based products or intrusion mouldings of polystyrene.

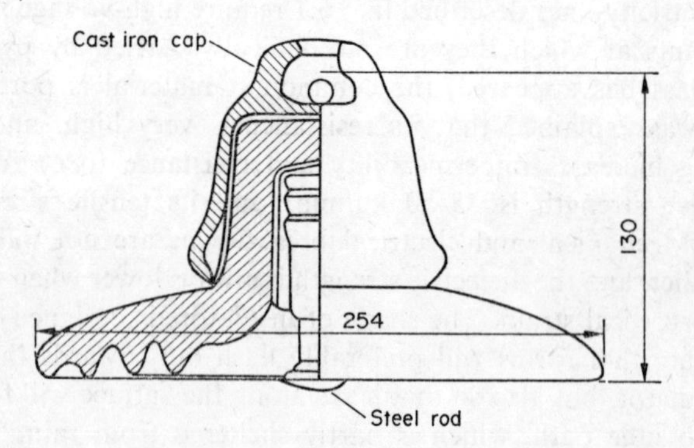

FIG. 10.1. Shape of "cap and rod" suspension insulators (based on Ailleret, 1945–6). Weight approximately 5 kg. Dimensions in millimetres.

Finally, a large number of electrotechnical insulators are used in static machinery (transformers) or revolving machinery (motors, alternators, and turboalternators), the circuits of which have to be insulated from one another, and from the magnetic plates and carcass, while ensuring compactness and resistance to temperature. Table 10.1 shows the various maximum temperatures permitted in continuous operation for windings, depending on the standardized category of the insulator used. Categories A, E or B are used in low- and medium-voltage a.c. machines, categories B or F in large synchronous machines, and categories F, H or higher in small low-voltage d.c. machines. In alternating machines from 500 to 6000 V, category A (cotton, silk, and paper impregnated with conventional organic varnishes) is used less and less, preference being given to category E (paper and polyethylene or polypropylene) and even, if possible, paper–mica tapes with a polyester, terephthalate or Mylar binder, or on a glass backing pre-impregnated with an epoxy resin, all in category B.

Table 10.1. Based on Leroy, 1964

Category	Y	A	E	B	F	H	C
Maximum temperature (°C)	90	105	120	130	155	180	>180

In large synchronous machines, conditions are much more exacting: maximum working temperature prevailing constantly over a working life of around 20 years, working voltage often attaining 20 kV and thus requiring low dielectric losses, major mechanical stresses on turbo-alternators 5 m long. These conditions rule out categories Y, A and even E, but they also exclude category H, the materials of which would not withstand mechanical strains so well. Insulation with sheets of conglomerated mica coated with asphalt has recently been replaced by a process for binding continuous micaceous tape with a synthetic resin which is thermosetting without solvent (Thermalastic process by Westinghouse). Figure 10.2 shows the comparative change in the loss factor tan δ in relation to voltage and temperature for the former process (micafolium–asphalt) and the new process (mica–polyester). This process involves composite mica, paper, glass, or Mylar tapes, bound either with a polyester resin of the bis-glycol-adipic-maleic type, which gives good mechanical properties, or with a styrene monomer, which ensures particularly low dielectric losses, copolymerized in the presence of benzoyl peroxide (Leroy, 1964).

In d.c. machines of up to 6000 V, used for instance in electrical traction on railways or in the navy, the need for maximum reduction of weight has led to a search for insulating materials that will keep their properties at high temperatures. Silicone resins are in almost general use, with final impregnation and polymerization by heating of the whole machine. Used in the form of glass–silicone or glass–mica–silicone

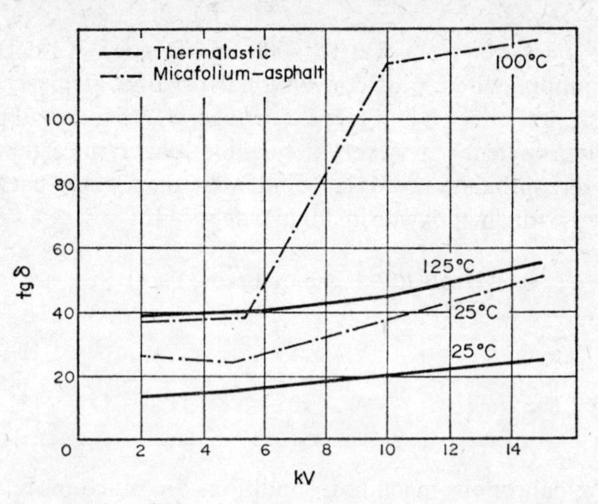

FIG. 10.2. Compared variation in the loss factor for the old and new insulating process for large synchronous machines in relation to voltage and temperature (based on Leroy, 1964).

Reprinted from "Dielectric Materials and Applications" by A. R. von Hippel by permission of The M.I.T. Press, Cambridge, Massachusetts.

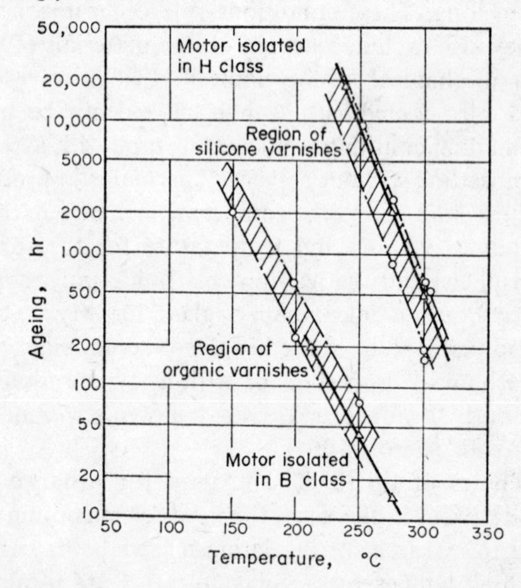

FIG. 10.3. Compared ageing of conventional organic varnishes (B class) and silicone varnishes (H class) in relation to temperature (based on Berberich, 1954).

Reprinted from "Dielectric Materials and Applications" by A. R. von Hippel by permission of The M.I.T. Press, Cambridge, Massachusetts tapes or laminates, they retain their qualities between −60 and +200°C, thus slightly exceeding the temperature limit for category H. Figure 10.3 shows some statistical results for the heat endurance of silicone varnishes compared with alkyde or phenolic resin-based varnishes. Samples used in these experiments consisted of 2/10 mm glass-fibre fabrics, hot-cleaned and impregnated with two coats of the varnish to be investigated. The duration of ageing corresponds to a drop of 50% in dielectric strength.

10.2. Electromechanical resonator

The use of quartz crystals as electromechanical *resonators* goes back a long time. It was seen in § 5.4 that the resonance of a piezoelectric material occurs when the reactance X in parallel with the capacity C_e of the equivalent circuit is nil. Figure 10.4 shows the composition of this reactance: mechanical rigidity C_m, mechanical mass L_m, and mechanical dampening R in series. Resonance and antiresonance correspond to two zero values of X. The quality factor Q is very high, around 10^5 for quartz, namely a hundred times greater than for normal resonating circuits at the same frequencies. Depending on its orientation,

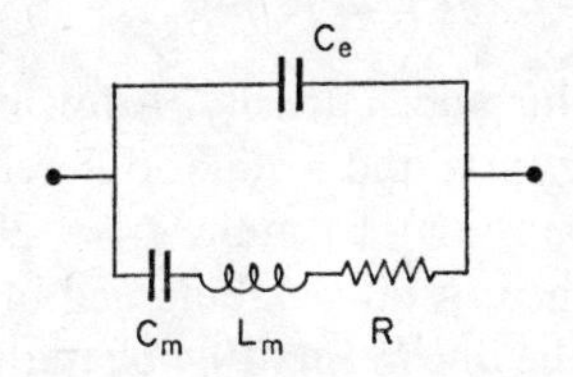

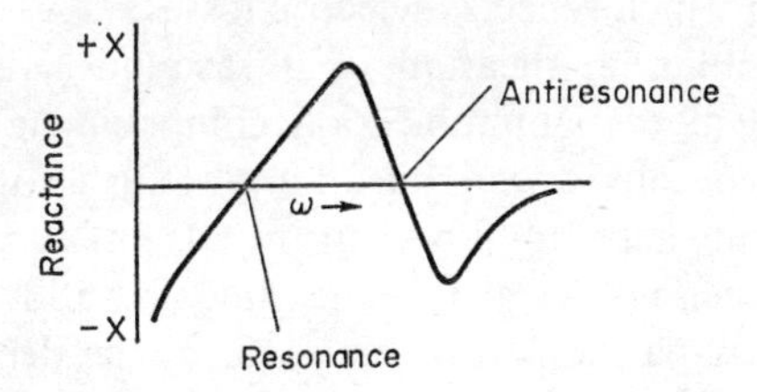

FIG. 10.4. Equivalent circuit and frequency response of a piezoelectric crystal near resonance (based on von Hippel, 1954).

Reprinted from "Dielectric Materials and Applications" by A. R. von Hippel by permission of The M.I.T. Press, Cambridge, Massachusetts.

quartz presents the section *X*, in which the electromechanical coupling coefficient is very high, and the section *GT*, in which the temperature coefficient of the piezoelectric coefficients is practically nil. Quartz is at present tending to be replaced by ceramic materials based on barium titanate powder, to which small amounts of dissolvents, diluents, or soluble oxides have been added. This method of preparation is more adaptable, but the titanate has high temperature coefficients, and dielectric losses increase very quickly from 200 V_{eff}/cm.

The main and longest-known use of this resonance is in the electromechanical piezoelectric transducer or *ultrasonic generator*, in other words which vibrates at frequencies higher than sounds audible to the human ear. Any piezoelectric crystal can be used, but when high ultrasonic power is needed only quartz and titanate ceramics are suitable (addition of lead to $BaTiO_3$ slightly raises the Curie point). For quartz, it is usually the *X* section that is chosen, vibrating in longitudinal mode. The length must be such that $\lambda = 2l/n$, where λ is the wavelength of the excited stationary wave and n an integer. The ultrasonic wave frequency is $f = v\lambda$, where v is the speed of sound in the crystal, namely

$$f = (Y/\rho)^{\frac{1}{2}}(2l/n),$$

where Y is Young's modulus and ρ density. Allowing for the values of piezoelectric constants in quartz and a slight difference between theory and practice, the semi-empirical formula $f = 2830l$ kHz is used (Anderson, 1964). When the crystal is submerged in a liquid, the equivalent circuit differs from the one in Fig. 10.4 by the addition of another series resistor, which represents the effect of the acoustic load, and a parallel resistor, which reflects dielectric losses.

The most familiar application of ultrasounds is the "sonar", an underwater system of communication comprising a transmitter and detector. This is because communication systems using electromagnetic waves can no longer be employed in the case of a submerged object. Listening to the echo of a signal in the same way allows obstacles, such as icebergs or submarines, to be detected, or the depth of the seabed recorded. The generator then has to be fed by a thyristor and function for only a few microseconds, so as not to interfere with reception of the echo. Ultrasounds are also used to prepare emulsions or precipitations, to measure thicknesses, check homogeneity, measure levels and flow-rates, to polish surfaces, weld plastics or sheets of metals, drill hard metals, etc. The power can easily attain 1 W/cm^2, in other words some 100 W for a mosaic of crystals connected in series or parallel. Since the laws of propagation of ultrasounds are fairly similar to those prevailing in optics, metal mirrors and plastic lenses can be constructed to focus them and concentrate energies of 500 W/cm^2.

Conversely, small vibrations can be converted into electrical signals, as in the case of microphones and record pick-up heads. Dielectric losses are of no importance in these applications, which, on the other hand, require high piezoelectric coefficients and good thermal stability. Rochelle salt is too hygroscopic, and has been superseded by ADP and KDP synthetic crystals and titanate and zirconate solid solutions, which have higher piezoelectric coefficients than pure titanates. The use of double elements on each side of a copper electrode allows a 1 V peak to be obtained for a normal recording.

A second application of electromechanical resonance is in piezo-electric *transformers*. A bar of piezoelectric ceramic is divided into two sections, totalling four electrodes (Fig. 10.5). A periodic voltage is

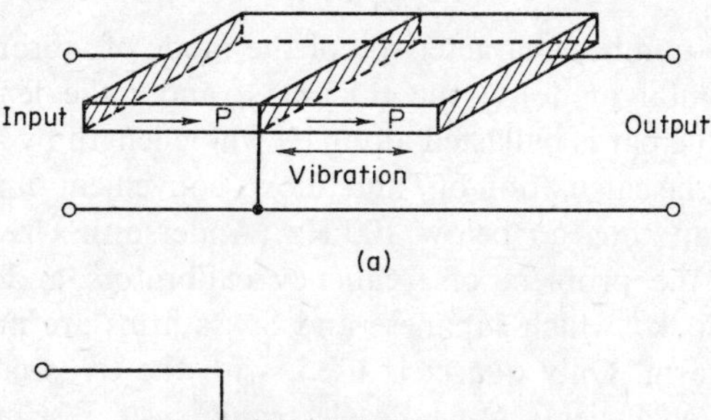

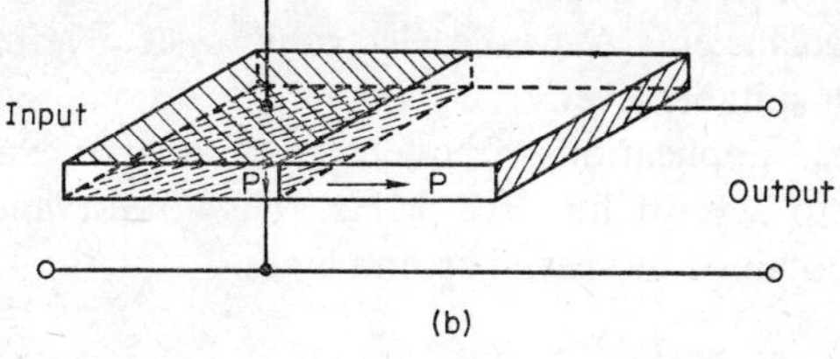

FIG. 10.5. Diagram of a piezoelectric transformer: (*a*) ring type, (*b*) transverse type (based on Anderson, 1964).

applied to the first section. If its length contains a whole number of half-wavelengths of the excited stationary wave, the mechanical oscillations immediately created there are transmitted to the adjoining section, where a secondary voltage thereupon appears, by direct piezoelectric effect. In the ring type (*a*), the ratio of the two voltages depends solely on the mechanical quality factor and electromechanical coupling coefficient, while in the transverse type (*b*), this ratio may be modified by the dimensions of the bar. In both cases, one has a high-voltage transformer, with high input and output impedances, generally used in high frequency. Gains are from 10 to 30, and efficiency low.

A final application of electromechanical resonance is in filtering certain frequencies, for which the high Q factor and stable resonance of piezoelectric components make them preferable to the electrical resonating circuits discussed in § 5.3. The simplest type of *filter* is obtained by producing longitudinal resonance in a long, thin bar consisting of two symmetrical sections separated by a common electrode, which links them and acts as an electrostatic screen. The mechanical resonance frequency is given by the equation quoted by Rayleigh for the vibrating deflection:

$$f = (Y/\rho)^{\frac{1}{2}}(m^2 t/4\pi l\sqrt{3}),$$

where m is a number characteristic of the mode of vibration (1.875 for the fundamental mode), t the thickness, and l the length. For low frequencies the bar is ballasted along its whole length by a lateral mass, which alters the calculation of f and allows convenient dimensions to be kept. One can thus go below 100 Hz (Anderson). Use as a filter is linked with the problem of frequency calibrators and the "crystal-controlled clock" which Japanese and Swiss firms are now offering in wristwatch form. Only quartz is used, with the *GT* section with zero temperature coefficient. A whole electronic system is used to reduce frequency to a suitable level.

For all these applications, frequencies of quartz resonators range from 1 kHz to several hundred MHz. The lowest and highest frequencies are obtained by special assemblages.

10.3. Condenser, amplifier

The dielectric qualities of an insulating material are shown mainly in the *condenser*, an electrical circuit component characterized by its capacity C, to which must be added, in the equivalent circuit, a resistor R' in series and a leak resistor, generally very high, in parallel on the whole system. It was seen, in § 5.3, that the tangent of the angle of loss is given by the ratio R'/X' if X' is the value of the reactance, here the capacitance $1/C\omega$, so that $\tan \delta = R'C\omega$. Table 10.2 shows the main types of condensers used, excluding liquid-dielectric condensers (electrolytic condensers).

Table 10.2

	Operating voltage (V)	Possible capacities
Ceramic part	12–5000	0.5 pF–50 nF
Glass tape	300–6000	1 pF–0.1 μF
Mica sheet	300–30,000	5 pF–0.5 μF
Plastic film	25–5000	20 pF–10 μF
Paper tape	160–10,000	1 nF–100 μF

The market for the condenser using an inorganic dielectric prepared by the ceramic technique is in the field of small capacities, since its initial breakthrough came from the use of ferroelectric materials to produce miniaturized components. The $BaTiO_3$ ceramic is not satisfactory because its relative dielectric constant or *permittivity* is less than 2000 at room temperature and very sensitive to temperature, and its loss factor is around 0.1. Use is accordingly made of solid solutions obtained by partial replacement of barium by strontium or of titanium by zirconium or tin. The first method reduces the Curie point to around room temperature, where permittivity reaches 5000, and the addition of a small amount of MgO can reduce losses to 0.02. The second method allows losses to be reduced, but also reduces permittivity. Iron and nickel oxides are also added. These improvements, together with the discovery of new materials, have allowed industrial production of bypass capacitors and check valves of 100 pF to 30 nF, in which losses drop to 0.01. The use in the same machines of non-ferroelectric oxides, such as TiO_2–MgO–ZrO_2 mixtures with a permittivity of 10–200, make it possible to produce condensers with much more stable characteristics in the picofarad range, with lower losses, of around 0.001. For all ceramic condensers, dielectric strength is a few kilovolts per millimetre (Suchet, 1954).

Condensers made with glass tapes are destined to replace those using sheets of pink Indian mica, available quantities of which are now restricted. A borosilicate glass with 4% of different oxides has been chosen. Both materials are used in thin sheets of 5–10 μm, alternating with sheets of aluminium, then coated with a plastic, after firing under pressure in the case of glass. Ranges of capacities obtained are comparable, but the dielectric strength of mica is naturally much greater. As for condensers using plastic or paper tapes, they have the advantage, for roughly the same thickness of dielectric material, of costing less; however, their temperature resistance is not so good. The plastics used are polystyrene, Teflon, kel-F, or Mylar. Metallization of tapes allows the size of condensers to be reduced. The final operation involves vacuum impregnation with a synthetic resin.

Mention must also be made of certain very thin-layer condensers recently produced by depositing dielectric material on a conducting support, in a vacuum of 10^{-5} mmHg, followed by metallization. The electrodes are usually of aluminium, and the dielectric material can be SiO (permittivity 5, losses 0.02 up to 1 MHz, dielectric strength 10^6 V/cm), LiF (permittivity 9.5), ZnS (permittivity 8), or MgF_2 (permittivity 5). A dielectric layer of 1000 Å has a breakdown voltage of approximately 10 V. This is too low for valve circuits, but is perfectly suitable for transistor circuits. In addition, the manufacturing technique fits excellently into the various operations needed to produce *integrated circuits*.

It was seen in § 5.4 that a condenser made with a ferroelectric material showed a hysteresis cycle below its Curie point. Its a.c. capacity accordingly depends, for small potential variations at its terminals, on the functioning point selected on the cycle. This point can be set as required by superimposing a d.c. polarization voltage on the a.c. voltage. For certain ferroelectric compositions, such as titanates containing stannic oxides, the relative capacity can range from 1, for zero polarization, to 0.1, for 150 V d.c. This gives a variable capacity, the value of which can be remote-controlled by altering its polarization. This property can be used to turn a ferroelectric condenser into an a.c. *amplifier*.

Figure 10.6 shows the simplest diagram for such an amplifier. The condenser is in series with the a.c. supply and the load. If the polarization voltage remains low compared with the voltage corresponding to saturation polarization, the point of functioning is on the steep part of the cycle, and the signal applied to the input will cause the capacity of the condenser to vary considerably. The a.c. traversing the load resistor depends on the value of the reactance $1/C\omega$. It will therefore be possible

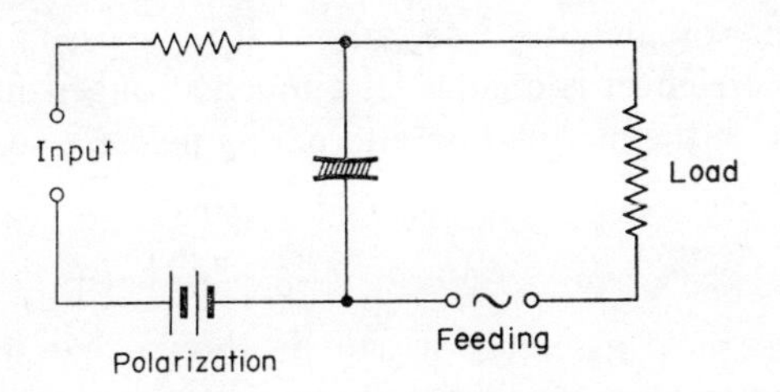

FIG. 10.6. Diagram of a simple dielectric amplifier (based on Anderson, 1964).

to obtain a power gain by a careful choice of the various impedances in the circuit. Efficiency can also be improved by moving from a single circuit to the bridge circuit, or even a resonating circuit. A total voltage gain of 10,000, for example, has been obtained in the frequency range 100 Hz to 10 kHz. Generalization of this type of application, however, is hindered by the effect of temperature on permittivity and by the high losses, despite its definite benefits (gain in size and weight compared with similar magnetic devices).

10.4. Light modulator

The birefringence phenomenon, described in § 8.4 in connection with magneto-optical effects, can also be due to *electro-optical effects*. Most liquids, when placed in an electric field E, take on the properties of a uniaxial crystal, the optical axis of which coincides with the direction of the field. Experiments have shown that the difference in the indices is expressed

$$n_e - n_0 = K\lambda E^2,$$

where K is Kerr's constant, approximately 10^{-7} e.m.u. It is agreed that the effect results from an orientation of the molecules of the liquid by the electric field, so that the value of the constant corresponds to the anisotropy of the molecule. The theory of molecular orientation shows that the sum of the three main indices (n_e+2n_0) is unchanged by the field, so that

$$(n_e-n)(n_0-n) = -2.$$

Experiments confirm this relation. The time needed to establish birefringence, approximately 10^{-8} sec, allows it to be used for modulation.

Since the Kerr effect is quadratic, a much stronger direct field E_0 is superimposed on the modulation alternating field E_1, so as to obtain a linear effect in E_1:

$$n_e-n_0 = K\lambda E^2 \sim K\lambda E_0^2+2K\lambda E_0 E_1.$$

In a transverse field assembly it can be shown that the modulation voltage needed to present a given phase shift is inversely proportional to the applied field E_0, often around 1–10 kV/cm. Initial research involved carbon sulphide and nitrobenzene. Recently, cubic centrosymmetrical crystals such as the perovskites $BaTiO_3$, $SrTiO_3$, and $KTaO_3$ have been used slightly above the ferroelectric Curie point. Although their ferroelectricity does not depend on the Kerr effect, this effect is in fact found to be considerably strengthened close to the Curie point. Below the Curie point, these substances are known to crystallize in a non-centrosymmetrical structure. The solid solutions $KTa_xNb_{1-x}O_3$ or KTN allow the Curie point to be lowered as far as room temperature with a permittivity of 10,000, and seem to hold promise.

Another electro-optical effect, the Pockels effect, consists of displacement of the ions of the crystal subjected to an electric field, different from the mechanical deformation resulting from the piezoelectric effect, which anyway has a neglible effect on birefringence at high frequencies, since mechanical distortion of the crystals can no longer follow the rapid variation in the field, and only birefringence resulting from the Pockels effect remains. The difference in the indices here is linear in relation to the electric field

$$n_e-n_0 = K'E.$$

If a light wave polarized along one of the neutral lines passes through a length l of crystal, it retains rectilinear polarization and undergoes a phase difference

$$2\varphi = 2\pi K' El/\lambda.$$

If E is a variable field, phase modulation is thereby obtained. First references to modulators built on this principle date from 1948, and 10 years later modulation frequencies of around 1 GHz were being attained. The longitudinal modulator, in which E is parallel to the direction of propagation, is to be distinguished from the transverse modulator, in which E is perpendicular to this direction. It is shown that the constant K' is proportional to the cube of the ordinary index.

Following ClO_3Na, α quartz and Rochelle salt, a large number of other crystals have shown a marked Pockels effect. First, there are the quadratic crystals ADP ($NH_4H_2PO_4$), KDP (KH_2PO_4), and deuterated KDP (KD_2PO_4), crystal growth of which is easy. The last of these has slightly higher dielectric losses, but tolerates lower operating voltages. Next come rhombohedral crystals $LiNbO_3$ and $LiTaO_3$, crystal growth of which is recent and which should contain only one ferroelectric domain. They are subject to optical damage at high light intensities. Finally, cubic crystals, which are non-birefringent in the absence of a field, still raise crystal growth problems (CuCl, ZnTe, CdTe), with the exception of chromium-doped GaAs, which can be used in the medium infrared range (Le Mézec, 1974).

Various types of Pockels-effect modulator have been conceived. In the one illustrated in Fig. 10.7 the difference in transit time between the two privileged vibrations of a fairly thick plate of a birefringent crystal is used to convert a polarization-state modulation into an amplitude modulation without loss of intensity. Other authors have converted the polarization-state modulation into a space modulation by using the Pockels effect in plates with no phase-lag in the absence of voltage and half-wave in the presence of voltage. They have thus obtained intermittent lateral translations of the beam. Variable-transmission modulators consisting of Kerr or Pockels cells placed between rectilinear polarizers are used for the photographic recording of sound on talking films, and experimental assemblies have allowed television pictures to be transmitted. A recent suggestion has involved a linear

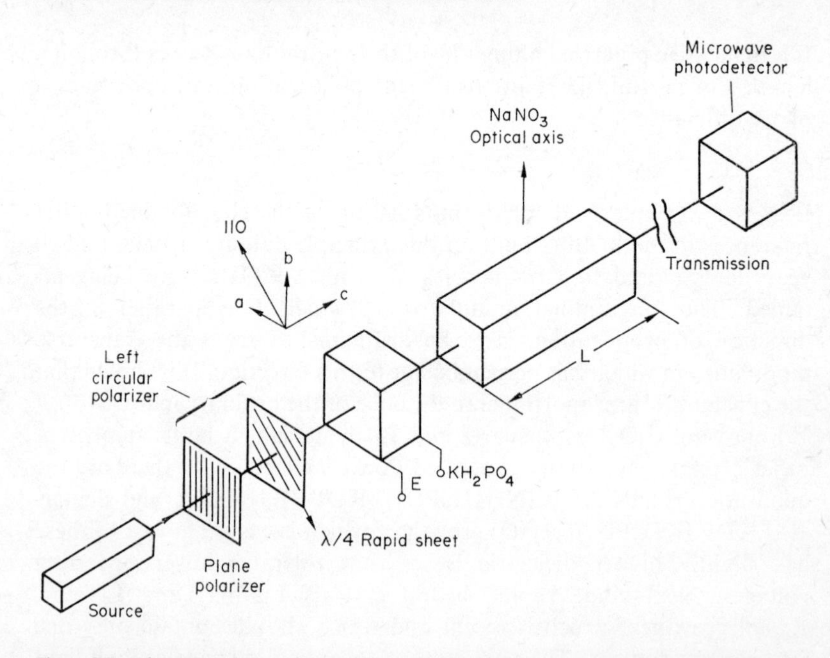

FIG. 10.7. Diagram of the conversion of a polarization-state modulation into amplitude modulation by passing through a thick birefringent plate (based on Buhrer, 1963).

modulator comprising a KDP crystal mounted in a laser cavity between the terminal mirror and a quartz plate with antireflective coating. The power of the laser is reduced from 1.2 to 0.4 mW by the presence of the crystal, and the frequency emitted can vary from ± 44 MHz with a sensitivity of 0.4 MHz/V to the voltage applied to the crystal (Suchet, 1971).

Birefringence resulting from electro-optical effects is not the only electrical phenomenon that can be used in a light modulator. A continuous, variable mechanical deformation, which also generates birefringence, can contribute to it. Effects caused by an alternating mechanical stress take on particular aspects at high frequencies, and the progressive or stationary acoustic waves created in a transparent solid by the piezoelectric effect constitute an effective diffraction lattice. Such *acousto-optical effects* have been used to obtain modulation by deflecting light. The most promising materials are melted SiO_2, $PbMoO_4$, and TeO_2 (Le Mézec, 1974).

10.5. Memory, electret

The storage of data in a *memory* is one of the main problems of our time, and although none of the solutions based on electrical conduction in insulating solids is at present used on a very wide scale, either in digital computers or in telephone exchanges, we feel that some account should be given of it. The basic component of a memory is a certain amount of material that can occupy several different states. Ten states are neither necessary nor desirable because of the risks of mistakes they would involve. In addition, immobilization of ten components to represent a figure in the decimal system would involve wastage of resources because, in the binary system, each of these components can represent a figure (0 or 1), so that together they can represent $2^{10}-1 = 1023$ different numbers. Only two stable states are accordingly needed, e.g. a positive or negative impulse, a closed or open relay, a conducting or non-conducting electron tube, a ferromagnetic or ferro-electric solid with two remanent states, a tunnel diode with two stable states, etc. The capacity of the memory is expressed in a number of these binary figures or *bits*.

Memories are initially classified not on the basis of the materials used in them but by their form of access, namely the way in which it is possible to feed in or remove data. *Sequential access*, which was used immediately after the Second World War, corresponds to the passing of all the information in the memory in an immutable order, each item occupying a given position. *Free access*, which is much more common now, involves a mass of information in bulk, each item of which can be reached directly by a selection system. In the first case, *access time* depends on the position of the information in the memory, while for the second it depends only on the nature of the physical effect used to produce the two stable states. It is convenient to use the following classification: slow memories above 1 msec (sequential access to a magnetic drum), medium from 1 msec to 10 μsec (sequential access to acoustic delay lines, free access to electrostatic tubes), high-speed from 10 to 0.1 μsec (free access to magnetic cores or thin layers), and ultra-high-speed below 0.1 μsec (free access to tunnel diodes, now being developed).

Quartz acoustic *delay lines*, which were used in certain sequential-access memories, make use of the low speed of sound in melted silica (4800 m/sec). Figure 10.8 illustrates their operation in diagrammatical form. The output signal is reinjected into the input after amplification and filtration so as to eliminate distortions. The sequence of positive or negative acoustic impulses thus circulates without stopping, and can be consulted at any time. Different ways of obtaining angled paths have been suggested to avoid the need for excessively large bars (4.8 m for 1 msec). It was seen, in § 9.5, that certain chalcogenide glasses could with advantage be used instead of melted silica for this application.

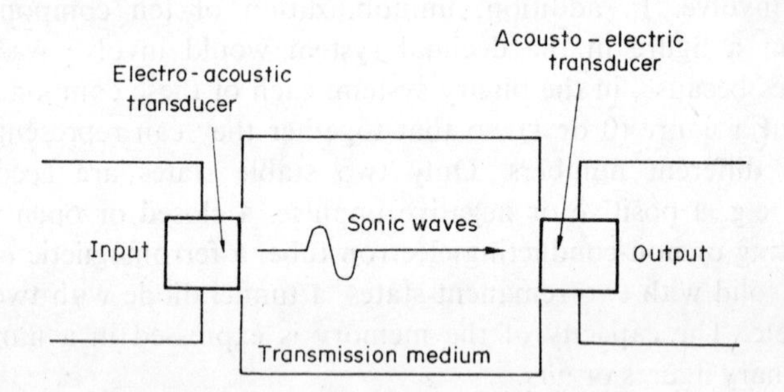

FIG. 10.8. Main parts of an acoustic delay line (based on Renwick, 1967).

Ferroelectric materials, like ferromagnetic materials, when they have a *rectangular hysteresis cycle*, are suitable for the construction of bistable components used in free-access memories. The simplest method consists of a thin ceramic plate of barium titanate, with parallel metallized lines on both sides, horizontal on one side and vertical on the other. The intersection of two lines thus forms a small ferroelectric condenser which can be charged with one polarity (figure 0) or the other (figure 1). However, the system has drawbacks compared with the corresponding magnetic systems: the cycles are less stable, and temperature sensitivity is greater.

Ferroelectric condensers, by the electrostatic charges they retain on each of their surfaces after being subjected to an electric field, can be regarded as the dielectric equivalent of permanent magnets. Substances capable of retaining permanent electric charges in this way are called *electrets*. They are closely connected with memories, several of which use their special properties. Section 5.4 explained how certain dielectric materials take on permanent charges in the absence of a field, by heating (pyroelectricity) or mechanical action (piezoelectricity), or by the application of an electric field (ferroelectricity). However, there are many other ways of preparing electrets. The commonest is heat treatment in a field, which was first applied successfully to different types of waxes, such as Carnauba wax. Thermoplastic materials heated in the softening zone, and then cooled in a field, show electret behaviour and optical anisotropy (Gross, 1964). Finally, Latour has recently (1972) mentioned polymerization of ethoxyline in the presence of a field.

As we have said, the use of electrets has been proposed in certain electrostatic memories (Williams cathodic tube in which the electronic beam charges a point on the screen, dielectric drum rotating beneath a brush electrode, charging of a plastic tape by a metal electrode, etc). They are also incorporated in electrostatic generators in the form of belts or discs carrying charges, and are used in duplicating, with the charged points of an electrostatic image fixing the grains of a fine, carbon powder. Air filters using them to fix small solid particles in suspension have also been proposed.

The combination of a flexible membrane electret and of an electrode, the surface of which is embossed, allows a capacity variable with the pressure to be realized and a sensibility of about, or even higher than, 1 mV/mb to be obtained (Alliot, 1974).

References

AILLERET, P. (1945–6) *Cours d'électrotechnique appliquée*, École Nat. Ponts et Chaussées, Paris.

ALLIOT, J-C. (1974) *La Recherche aérospatiale* (Fr.) **2**, 95.

ANDERSON, J. C. (1964) *Dielectrics*, Chapman & Hall, London (French translation, Dunod, Paris, 1966).

BERBERICH, L. J. (1954) *Dielectric Materials and Applications* (ed. A. R. von Hippel), p. 190, Wiley, New York (French translation, Dunod, Paris, 1961).

BILLARD, J. (1966) *Rev. Phys. Appl.* **1**, 311.

BUHRER, C. F. (1963) *Proc. IEEE* **51**, 1151.

GROSS, B. (1964) *Charge Storage in Solid Dielectrics*, Elsevier, Amsterdam.

LATOUR, M. (1972) *C.R. Acad. Sci. Paris* **274 B**, 874.

LE MÉZEC, J. (1974) *Séminaires de chimie de l'état solide* (ed. J. P. Suchet), **7**, 89, Masson, Paris.

LEROY, L. (1964) *Industrie des plastiques modernes* (Fr.) **16,** no. 4 (October issue).
RENWICK, W. (1954) *Digital Storage Systems*, Spon, London (French translation, Eyrolles, Paris, 1967).
SUCHET, J. P. (1954) *Onde électr*. (Fr.) **34,** 460.
SUCHET, J. P. (1971) *Crystal Chemistry and Semiconduction in Transition Metal Binary Compounds*, Academic Press, New York.
VON HIPPEL, A. R. (1954) *Dielectric Materials and Applications* (ed. A. R. von Hippel), p. 1, Wiley, New York.

Guide to Recent Books

(abstracted in *Chemical Abstracts* between Jan. 1, 1970 and Dec. 31, 1973)

GENERAL

Radiation Effects Design Handbook (NASA Contractor Report 1787), NTIS, Springfield, Va., USA, 1971.

Handbook of Electronic Materials, IFI/Plenum, New York, NY, USA, 1972.

Annual Book of ASTM Standards, American Society for Testing and Materials, Philadelphia, Pa., USA, 1973.

PART ONE: PHYSICOCHEMICAL BASES

Conductors

BRAUER, W., *Einführung in die Elektronentheorie der Metalle* (*Introduction to the Electron Theory of Metals*), 2nd edn., Geest und Portig, Leipzig, E. Ger., 1972 (in German).

BARNARD, R. D., *Thermoelectricity in Metals and Alloys*, Taylor and Francis, London, Engl., 1972.

Superconductors

ROSE-INNES, A. C. and RHODERICK, E. H., *Introduction to Superconductivity* (International Monographs on Solid State Physics, Vol. 6), Pergamon, New York, NY, USA, 1969.

FISHLOCK, D., *A Guide to Superconductivity*, Elsevier, New York, NY, USA, 1969.

MÜLLER, J., *Phenomenological Aspects of Superconductivity*, CERN, Geneva, Switz., 1969.

ROBERTS, B. W., *Superconductive Materials and Some of their Properties*, GPO, Washington, DC, USA, 1969.

CONRAD, P., *Superconducting Materials*, Noyes Data Corp., Park Ridge, NJ, USA, 1970.

WILLIAMS, J. E. C., *Superconductivity and its Applications*, Pion, London, Engl., 1970.

FIRTH, I. M., *Superconductivity* (M and B Monographs, EE/9), Mills and Boon, London, Engl., 1972.

BOL'SHANINA, M. A., *Sverkhprovodimost'* (*Superconductivity*), Izd. Tomsk. Univ., Tomsk, USSR, 1972 (in Russian).

Conventional Semiconductors

GORYUNOVA, N. A., *Khimiya almazopodobnykh poluprovodnikov* (*The Chemistry of Diamondlike Semiconductors*), Izd. Leningrad. Gos. Univ., Leningrad, USSR, 1963 (in Russian). English translation: M.I.T. Press, Cambridge, Mass., U.S.A., and Chapman & Hall, London, Engl., 1965. German translation: *Halbleiter mit diamantähnlicher Struktur*, Teubner, Leipzig, E. Ger., 1971.

BERGER, L. I. and PROCHUKHAN, V. D., Printed in USSR (in Russian). English translation: *Ternary Diamondlike Semiconductors*, Consultants, New York, NY, USA, 1969.

RAVICH, YU. I., EFIMOVA, B. A., SMIRNOV, I. A., *Metody issledovaniya poluprovodnikov v primenii k khal'kogenidam svintsa* (*PbTe, PbSe i PbS*) (*Methods of Investigation of Semiconductors Applied to Lead Chalcogenides* (*PbTe, PbSe, and PbS*)), Nauka, Moscow, USSR, 1968 (in Russian). English translation: *Semiconducting Lead Chalcogenides* (Monographs in Semiconductor Physics, Vol. 5), Plenum, New York, NY, USA, 1970.

REGEL, A. R., *Metody issledovaniya termoelektricheskikh svoistv poluprovodnikov* (*Methods of Investigation of the Thermoelectric Properties of Semiconductors*), Atomizdat, Moscow, USSR, 1969 (in Russian).

SHALABUTOV, Yu. K., *Vvedenie v fiziku poluprovodnikov* (*Introduction to the Physics of Semiconductors*), Nauka, Leningrad Otd., Leningrad, USSR, 1969 (in Russian).

SIROTA, N. N., *Khimicheskaya svyaz' v poluprovodnikakh* (*Chemical Bonding in Semiconductors*), Nauka i Tekhnika, Minsk, Bel. SSR, 1969 (in Russian). English translation: *Chemical Bonds in Solids*, Vols. 1–4 (Proceedings of the International Symposium on Chemical Bonds in Semiconducting Crystals, held in Minsk, USSR, in 1967), Consultants, New York, NY, USA, 1972.

KANE, P. F. and LARRABEE, G. B., *Characterization of Semiconductor Materials*, McGraw-Hill, New York, NY, USA, 1970.

MIRONOV, K. E. and VASIL'EVA, I. G., *Khimiya fosfidov s poluprovodnikovymi svoistvami* (*Chemistry of Phosphides with Semiconductor Properties*), Nauka, Sib. Otd., Novosibirsk, USSR, 1970 (in Russian).

MÜLLER, R., *Grundlagen der Halbleiter-Elektronik* (*Halbleiter-Elektronik*, Bd. 1) (*Principles of Semiconductor Electronics* (*Semiconductor Electronics*, Vol. 1)), Springer, New York, NY, USA, 1971 (in German).

CLOSE, K. and YARWOOD, J., *An Introduction to Semiconductors*, Heinemann Ed. Books, London, Engl., 1971.

BYLANDER, E. G., *Materials for Semiconductor Functions* (Hayden Series in Materials for Electrical and Electronic Design), Hayden, New York, NY, USA, 1971.

COHEN, M. M., *Introduction to the Quantum Theory of Semiconductors*, Gordon & Breach, New York, NY, USA, 1972.

OKHOTIN, A. S., PUSHKARSKII, A. S. and GORBACHEV, V. V., *Teplofizicheskie svoistva poluprovodnikov* (*Thermophysical Properties of Semiconductors*), Atomizdat, Moscow, USSR, 1972 (in Russian).

PHILLIPS, J. C., *Bonds and Bands in Semiconductors*, Academic, New York, NY, USA, 1973.

Other Semiconductors

BOGUSLAVSKII, L. I. and VANNIKOV, A. V., Printed in USSR (in Russian). English translation: *Organic Semiconductors and Biopolymers* (Monographs in Semiconductor Physics, Vol. 6), Plenum, New York, NY, USA, 1970.

MYULLER, R. L., *Elektroprovodnost' stekloobraznykh vestchetv* (*Electrical Conductivity of Vitreous Substances*), Izd. Leningrad. Gos. Univ., Leningrad, USSR, 1968 (in Russian). English translation: Consultants, New York, NY, USA, 1971.

JARZEBSKI, Z. M. and MROWEC, S., *Polprzewodniki tlenkowe* (*Nowa technika*, 81) (*Oxide Semiconductors* (*New Techniques*, No. 81)), WNT, Warsaw, Pol., 1969 (in Polish).

DOREMUS, W., *Semiconductor Effects in Amorphous Solids*, North-Holland, Amsterdam, Neth., 1970.

DULOV, A. A. and SLINKIN, A. A., *Organicheskie poluprovodniki. Polimeri s sopyrazhennymi svyazyami* (*Organic Semiconductors. Polymers with Conjugate Bonds*), Nauka, Moscow, USSR, 1970 (in Russian).

SUCHET, J. P., *Crystal Chemistry and Semiconduction in Transition Metal Binary Compounds*, Academic, New York, NY, USA, 1971.

ADLER, D., *Amorphous Semiconductors* (CRC Monotopics Series), CRC Press, Cleveland, Ohio, USA, 1971.

X, *Fundamentals of Amorphous Semiconductors*, National Academy of Sciences, Washington, DC, USA, 1972.

GOL'TSMAN, B. M., KUDINOV, V. A. and SMIRNOV, I. A., *Poluprovodnidovye termoelektricheskie materialy na osnove Bi_2Te_3* (*Fizika poluprovodnikov i poluprovodnikovykh priborov*) (*Semiconductive Thermoelectric Materials Based on Bi_2Te_3* (*Physics of Semiconductors and Semiconductor Devices*)), Nauka, Moscow, USSR, 1972 (in Russian).

KOFSTAD, P., *Nonstoichiometry, Diffusion, and Electrical Conductivity in Binary Metal Oxides* (Wiley Series on the Science and Technology of Materials), Wiley–Interscience, New York, NY, USA, 1972.

BORISOVA, Z. U., *Khimiya stekloobrazhykh poluprovodnikov* (*Chemistry of Glassy Semiconductors*), Izd. Leningrad. Univ., Leningrad, USSR, 1972 (in Russian).

Insulators

FESENKO, E. G., *Polyarizatsiya p'ezokeramiki* (*Polarizing Piezoceramics*), Izd. Rostovsk. Univ., Rostov-on-Don, USSR, 1968 (in Russian).

HILL, N. E., VAUGHA, W. E., PRICE, A. H. and DAVIES, M., *Dielectric Properties and Molecular Behavior* (The van Nostrand Series in Physical Chemistry), van Nostrand–Reinhold, New York, NY, USA, 1969.

HEDVIG, P., *Elektromos vezetes es polarizacio muanyagokban* (*Electric Conduction and Polarization in Plastics*), Akad. Kiado, Budapest, Hung., 1969 (in Hungarian).

GRINDLAY, J., *An Introduction to the Phenomenological Theory of Ferroelectricity* (International Series of Monographs in Natural Philosophy, Vol. 26), Pergamon, New York, NY, USA, 1970.

MAIOFIS, I. M., *Khimiya dielektrikov* (*Chemistry of Dielectrics*), Vysshaya Shkola, Moscow, USSR, 1970 (in Russian).

SMOKE, E. J., *Inorganic Dielectrics Research: A History of Twenty-Three Years of Ceramic Dielectric Research* (Engineering Research Bulletin, No. 50), Rutgers Univ., New Brunswick, NJ, USA, 1970.

JAFFE, B., COOK, W. R., JR. and JAFFE, H., *Piezoelectric Ceramics* (Nonmetallic Solids, No. 3), Academic, New York, NY, USA, 1971.

SMAZHEVSKAYA, E. G. and FEL'DMAN, N. B., *P'ezoelektricheskaya keramika* (*Piezoelectric Ceramics*), Sovetskoe Radio, Moscow, USSR, 1971 (in Russian).

FESENKO, E. G., *Semeistvo perovskita i segnetoelektrichestvo* (*Perovskite Family and Ferroelectricity*), Atomizdat, Moscow, USSR, 1972 (in Russian).

Miscellaneous

FIKS, V. B., *Ionnaya provodimost' v metallakh i poluprovodnikakh* (*Ionic Conductivity of Metals and Semiconductors*), Nauka, Moscow, USSR, 1969 (in Russian).

GLAZOV, V. M. and VIGDOROVICH, V. N., *Mikrotverdost' metallov i poluprovodnikov*, Izd. 2–e (*Microhardness of Metals and Semiconductors*, 2nd edn.), Metallurgiya, Moscow, USSR, 1969 (in Russian).

LITTLE, W. A., *Organic Superconductors* (Polymer Symposia, No. 29), Interscience, New York, NY, USA, 1970.

KISLEV, V. F., *Poverkhostnye yavleniya v poluprovodnikakh i dielektrikakh* (*Surface Phenomena in Semiconductors and Dielectrics*), Nauka, Moscow, USSR, 1970 (in Russian).

VIJH, A. K., *Electrochemistry of Metals and Semiconductors: The Application of Solid State Science to Electrochemical Phenomena*, Dekker, New York, NY, USA, 1973.

PART TWO: POSSIBLE APPLICATIONS

Conductor Technology

DE RENZO, D. J., *Wire Coatings*, Noyes Data Corp., Park Ridge, NJ, USA, 1971.

Superconductor Technology and Devices

SAVITSKII, E. M., *et al.*, *Metallovedenie sverkhprovodyashchikh materialov* (*Metallurgy of Superconducting Materials*), Nauka, Moscow, USSR, 1969 (in Russian).

SAVITSKII, E. M. and BARON, V. V., *Fiziko-khimiya, metallovedenie i metallofizika sverkhprovodnikov* (*Physicochemistry, Physical Metallurgy, and Metal Physics of Superconductors*), Nauka, Moscow, USSR, 1969 (in Russian). English translation: *Physics and Metallurgy of Superconductors*, Plenum, New York, NY, USA, 1970.

KULIK, I. O. and YANSON, I. K., *Effekt dzhozefsona v sverkhprovodyashchikh tunnel'nykh strukturakh* (*The Josephson Effect in Superconductive Tunnel Structures*), Nauka, Moscow, USSR, 1970 (in Russian). English translation: Wiley, New York, NY, USA, 1972.

WILLIAMS, J. E. C., *Superconductivity and its Applications*, Pion, London, Engl., 1970

IVANOV, O. S., RAEVSKII, I. I. and STEPANOV, N. V., *Sverkhprovodyashchie splavy sistemy niobii–titan–tsirkonii–gafnii* (*Superconducting Alloys of the System Niobium–Titanium–Zirconium–Hafnium*), Nauka, Moscow, USSR, 1971 (in Russian).

PETLEY, B. W., *An Introduction to the Josephson Effects* (M and B Technical Library TL/EE/2), Mills & Boon, London, Engl., 1971.

SOLYMAR, L., *Superconductive Tunneling and Applications*, Chapman & Hall, London, Engl., 1972.

Semiconductor Technology

SCHWARTZ, B., *Ohmic Contacts to Semiconductors*, Electrochemical Society, New York, NY, USA, 1969.

SITTIG, M., *Manufacture of Semiconductor Compounds*, Noyes Development Corp., Park Ridge, NJ, USA, 1969.

SITTIG, M., *Pure Chemical Elements for Semiconductors* (Electronics Materials Review, No. 1), Noyes Development Corp., Park Ridge, NJ, USA, 1969.

Sittig, M., *Semiconductor Crystal Manufacture* (Electronics Materials Review, No. 3), Noyes Data Corp., Park Ridge, NJ, USA, 1969.

Vigdorovich, V. N., *Ochista metallov i poluprovodnikov kristallizatsiei* (*Refining of Metals and Semiconductors by Crystallization*), Metallurgiya, Moscow, USSR, 1969 (in Russian). English translation (Freund Materials Science and Engineering Series), Freund, Holon, Israël, 1971.

Pankrat'ev, E. M., Ryumin, V. P. and Shchelkina, N. P., *Tekhnologiya poluprovodnikovykh sloev dvuokisi olova* (*Technology of Stannic Oxide Semiconductor Films*), Energiya, Moscow, USSR, 1969 (in Russian).

Klochkov, V. P. and Svechnikov, S. V., *Tonkoplivkovi napivprovidnikovi materialy v mikroelektronitsi* (*Thin-film Semiconductor Materials in Microelectronics*), Tekhnika, Kiev, Ukr. SSR, 1969 (in Ukrainian).

Mayer, J. W., Eriksson, L. and Davies, J. A., *Ion Implantation in Semiconductors, Silicon and Germanium*, Academic, New York, NY, USA, 1970.

Wieder, H. H., *Intermetallic Semiconducting Films* (International Series of Monographs in Semiconductors, Vol. 10), Pergamon, New York, NY, USA, 1970.

Medvedev, S. A., *Vvedenie v teknologiyu poluprovodnikovykh materialov* (*Introduction to the Technology of Semiconductor Materials*), Vyssh. shkola, Moscow, USSR, 1970 (in Russian).

Krasulin, Yu. L., *Vzaimodeistvie metalla s poluprovodnikom v tverdoi faze* (*Solid Phase Reactions of Metals with Semiconductors*), Nauka, Moscow, USSR, 1971 (in Russian).

Orton, J. W., *Material for the Gunn Effect* (M and B Monograph EE/3), Mills & Boon, London, Engl., 1971.

Connolly, T. F., *Solid-state Physics Literature Guides*, Vol. 2, *Semiconductors. Preparation, Crystal Growth, and Selected Properties*, IFI/Plenum, New York, NY, USA, 1972.

Nashel'skii, A. Ya., *Tekhnologiya poluprovodnikovykh materialov* (*Technology of Semiconductor Materials*), Metallurgiya, Moscow, USSR, 1972 (in Russian).

Semiconductor Devices

Nosov, Yu. R., Printed in USSR (in Russian). English translation: *Switching in Semiconductor Diodes* (Monographs in Semiconductor Physics, Vol. 4), Plenum, New York, NY, USA, 1969.

Sze, S. M., *Physics of Semiconductor Devices*, Wiley–Interscience, New York, NY, USA, 1969.

Weiss, H., *Structure and Application of Galvanomagnetic Devices* (International Series of Monographs on Semiconductors, Vol. 8), Pergamon, New York, NY, USA, 1969.

Hartnagel, H., *Semiconductor Plasma Instabilities, Including Gunn Effect and Avalanche Oscillations*, Elsevier, New York, NY, USA, 1969.

Milnes, A. G. and Feucht, D. L., *Heterojunctions and Metal–Semiconductor Junctions*, Academic, New York, NY, USA, 1972.

Campbell, A. M. and Evetts, J. E., *Critical Currents in Semiconductors*, Harper & Row, New York, NY, USA, 1972.

Insulator Technology and Devices

Vrantny, F., *Thin Film Dielectrics*, Electrochemical Society, New York, NY, USA, 1969.

Kramarov, O. P., *P'ezoelektricheskie materialy i preobrazovateli* (*Piezoelectric Materials and Transducers*), Izd. Rostovsk. Univ., Rostov-on-Don, USSR, 1969 (in Russian).

Ponomarev, L. T., *Eskaponovaya elektricheskaya izolyatsiya* (*Eskapon Electric Insulation*), Energiya, Leningrad. Otd., Leningrad, USSR, 1969 (in Russian).

Koritskii, Yu. V., *Proizvodstvo, svoistva i primenenie elektroizolyatsionnykh tsellyuloznykh bumag i kartonov* (*Production, Properties, and Use of Electric-insulating Cellulose Papers and Paperboards*), Energiya, Moscow, USSR, 1970 (in Russian).

Macdonald, J., *Metal–Dielectric Multilayers* (Monographs on Applied Optics, No. 4), Hilger, London, Engl., 1971.

Renne, V. T., *Plenochnye kondensatory s organicheskim sinteticheskim dielektrikom*, Izd. 2–e (*Film Capacitors with Organic Synthetic Dielectrics*, 2nd edn.), Energiya, Leningrad. Otd., Leningrad, USSR, 1971 (in Russian).

Ponomarenko, V. D., Rozdova, R. A., Ainshtein, R. G. and Goryacheva, G. A., *Stekloemalevye i steklokeramicheskie kondensatory* (*Glass–Enamel and Glass–Ceramic Capacitors*), Energiya, Moscow, USSR, 1972 (in Russian).

Kostyukov, N. S., Kharitonov, F. Ya. and Antonova, N. P., *Radiatsionnaya i korrozionnaya stoikost' elektrokeramiki* (*Radiation Resistance and Corrosion Resistance of Electrical Ceramics*), Atomizdat, Moscow, USSR, 1973 (in Russian).

Kan, K. N., Nikolaevich, A. F. and Shannikov, V. M., *Mekhanicheskaya prochnost' epoksidnoi izolyatsii* (*Mechanical Strength of Epoxy Insulation*), Energiya, Leningrad. Otd., Leningrad, USSR, 1973 (in Russian).

Miscellaneous

Miller, L. F., *Thick Film Technology and Chip Joining* (*Processes and Materials in Electronics*, Vol. 1), Gordon & Breach, New York, NY, USA, 1972.

Wilson, R. G. and Brewer, G. R., *Ion Beams with Applications to Ion Implantation*, Wiley–Interscience, New York, NY, USA, 1973.

FOR PHYSICISTS ONLY

Superconductors

Kresin, V. Z., *Sverkhprovodimost' i sverkhtekuchest'* (*Superconductivity and Superfluidity*), Prosveshchenie, Moscow, USSR, 1968 (in Russian).

Wallace, P. R., *Superconductivity*, Vols. 1 and 2, Gordon & Breach, New York, NY, USA, 1969.

Galasiewicz, Z. M., *Superconductivity and Quantum Fluids* (International Series of Monographs in Natural Philosophy, Vol. 29), Pergamon, New York, NY, USA, 1970.

Taylor, A. W. B., *Superconductivity*, Wykeham, London, Engl., 1970.

Williams, J. E. C., *Superconductivity and its Applications*, Pion, London, Engl., 1970.

Buckel, W., *Supraleitung. Grundlagen und Anwendungen* (*Superconductivity. Principles and Applications*), Physik Verlag, Weinheim, Ger., 1972 (in German).

Geilikman, B. T. and Kresin, V. Z., *Kineticheskie i nestatsionarnye yavleniya v sverkhprovodnikakh* (*Kinetic and Nonsteady-state Phenomena in Superconductors*), Nauka, Moscow, USSR, 1972 (in Russian).

Semiconductors

GREIG, D., *Electrons in Metals and Semiconductors*, McGraw-Hill, New York, NY, USA, 1969.

KALVENAS, S. B., *Actual'nye voprosy fiziki poluprovodnikov i poluprovodnikovykh priborov* (*Main Problems in Physics of Semiconductors and Semiconductor Devices*), Inst. fiz. poluprovodnikov akad. nauk Litov. SSR, Vil'nus, Lit. SSR, 1969 (in Russian).

KIREEV, P. S., *Fizika poluprovodnikov* (*Physics of Semiconductors*), Vysshaya shkola, Moscow, USSR, 1969 (in Russian).

MOTT, N. F., *Amorphous and Liquid Semiconductors*, North-Holland, Amsterdam, Neth., 1970.

KELLER, S. P., HENSEL, J. C. and STERN, F., *Physics of Semiconductors* (CONF-700801), Nat. Tech. Inf. Serv., Springfield, Va., USA, 1970.

TAVERNIER, J. and CALECKI, D., *Introduction aux phénomènes de transport linéaires dans les semiconducteurs* (*Introduction to Linear Transport Phenomena in Semiconductors*), Masson, Paris, Fr., 1970 (in French).

SHARMA, B. L., *Diffusion in Semiconductors*, Trans. Tech., Rocky River, Ohio, USA, 1970.

PANKOVE, J. I., *Optical Processes in Semiconductors*, Prentice-Hall, Englewood Cliffs, NJ, USA, 1971.

MATARE, H. F., *Defect Electronics in Semiconductors*, Interscience, New York, NY, USA, 1971.

BAINHAM, A. C. and BOARDMAN, A. D., *Plasma Effects in Semiconductors*, Taylor & Francis, London, Engl., 1971.

WOLF, H. F., *Semiconductors*, Interscience, New York, NY, USA, 1971.

RZHANOV, A. V., *Elektronnye protsessy na poverkhnosti poluprovodnikov* (*Fizika poluprovodnikov i poluprovodnikovykh priborov*) (*Electron Processes on the Surface of Semiconductors* (*Physics of Semiconductors and Semiconductor Devices*)), Nauka, Moscow, USSR, 1971 (in Russian).

SHALIMOVA, K. V., *Fizika poluprovodnikov* (*Physics of Semiconductors*), Energiya, Moscow, USSR, 1971 (in Russian).

KARAGEORGII-ALKALAEV, P. M. and LEIDERMAN, A. YU., *Glubokie primesnye urovni v shirokozonnykh poluprovodnikakh* (*Deep Impurity Levels in Broad-band Semiconductors*), Fan, Tashkent, Uzb. SSR, 1971 (in Russian).

KONOPLEVA, R. F., LITVINOV, V. L. and UKHIN, N. A., *Osobennosti radiatsionnogo povrezhdeniya poluprovodnikov chastitsami vysokikh energii* (*Characteristics of Radiation Damage of Semiconductors by High-energy Particles*), Atomizdat, Moscow, USSR, 1971 (in Russian).

MOSS, T. S., BURREL, G. J. and ELLIS, B., *Semiconductor Optoelectronics*, Butterworths, London, Engl., 1972.

MOGILEVSKII, B. M. and CHUDNOVSKII, A. F., *Teploprovodnost' poluprovodnikov* (*Fizika poluprovodnikov i poluprovodnikovykh priborov*) (*Thermal Conductivity of Semiconductors* (*Physics of Semiconductors and Semiconductor Devices*)), Nauka, Moscow, USSR, 1972 (in Russian).

BIR, G. L. and PIKUS, G. E., *Simmetriya i deformatsionnye effekty v poluprovodnikakh* (*Symmetry and Deformation Effects in Semiconductors*), Nauka, Moscow, USSR, 1972 (in Russian).

BOLTAKS, B. I., *Diffuziya i tochechnye defekty v poluprovodnikakh* (*Diffusion and Point Defects in Semiconductors*), Nauka, Leningrad. Otd., Leningrad, USSR, 1972 (in Russian).

BONCH-BRUEVICH, V. L., ZVYAGIN, I. P. and MIRONOV, A. G., *Domennaya elektricheskaya neustoichivost' v poluprovodnikakh* (*Fizika poluprovodnikov i poluprovodnikovykh priborov*) (*Domain Electrical Instability in Semiconductors* (*Physics of Semiconductors and Semiconductor Devices*)), Nauka, Moscow, USSR, 1972 (in Russian).

NAG, B. R., *Theory of Electrical Transport in Semiconductors* (International Series of Monographs in the Science of the Solid State, Vol. 3), Pergamon, Oxford, Engl., 1972.

TSIDIL'KOVSKII, I. M., *Elektrony i dyki v poluprovodnikakh. Energiticheskii spektr i dinamika* (*Electrons and Holes in Semiconductors. Energy Spectrum and Dynamics*), Nauka, Moscow, USSR, 1972 (in Russian).

KAIDANOV, V. I., ERASOVA, N. A. and ZHITINSKAYA, M. K., *Kineticheskie yavleniya v poluprovodnikakh*, Ch. 1 (*Kinetic Phenomena in Semiconductors*, Pt. 1), Leningrad. Politekh. Inst., Leningrad, USSR, 1972 (in Russian).

MILNES, A. G., *Deep Impurities in Semiconductors*, Wiley–Interscience, New York, NY, USA, 1973.

Insulators

ZHELUDEV, I. S., *Fizika kristallicheskikh dielektrikov* (*Physics of Crystalline Dielectrics*), Nauka, Moscow, USSR, 1968 (in Russian). English translation: Vols. 1 and 2, Plenum, New York, NY, USA, 1971.

SMOLENSKII, G. A. and KRAINIK, N. N., *Segnetoelektriki i Antisegnetoelektriki* (*Ferroelectrics and Antiferroelectrics*), Nauka, Moscow, USSR, 1968 (in Russian). German translation: *Ferroelektrika und Antiferroelektrika* (Mathematisch-Naturwissenschaftliche Bibliothek, Nr. 50), Teubner, Leipzig, E. Ger., 1972 (in German).

PIEKARA, A., *Fizyka dielektrykow i radiospektroskopia*, 1 (*Physics of Dielectrics and Radiospectroscopy*, Vol. 1), PWN, Poznan, Pol., 1969 (in Polish).

GUREVITCH, V. M., *Elektroprovodnost' segnetoelektrikov* (*Electroconductivity of Ferroelectrics*), Izd. Standart., Moscow, USSR, 1969 (in Russian).

MASHKOVICH, M. D., *Elektricheskie svoistva neorganicheskikh dielektrikov v diapazone SVCh* (*Electric Properties of Inorganic Dielectrics in the UHF Band*), Sov. Radio, Moscow, USSR, 1969 (in Russian).

ZAKY, A. A. and HAWLEY, R., *Dielectric Solids* (Solid State Physics), Dover, New York, NY, USA, 1970.

CONNOLLY, T. F. and TURNER, E., *Ferroelectric Materials and Ferroelectricity* (Solid State Physics Literature Guides, Vol. 1), IFI/Plenum, New York, NY, USA, 1970.

SMOLENSKII, G. A., BOKOV, V. A., ISUPOV, V. A., KRAINIK, N. N., PASYNKOV, R. E. and SHUR, M. S., *Segnetoelektriki i antisegnetoelektriki* (*Ferroelectrics and Antiferroelectrics*), Nauka, Leningrad. Otd., Leningrad, USSR, 1971 (in Russian).

KHOLODENKO, L. P., *Termodinamicheskaya teoriya segnetoelektrikov tipa titanata bariya* (*Thermodynamic Theory of Barium Titanate-type Ferroelectrics*), Zinatne, Riga, Latv. SSR, 1971 (in Russian).

STATISTICS (original language and place of publication)

ENGLISH	64	USA: New York, NY (34); Park Ridge, NJ (5); Springfield, Va. (2); Washington, DC (2); Cleveland, Ohio (1); Englewood Cliffs, NJ (1); New Brunswick, NJ (1); Philadelphia, Pa. (1); Rocky River, Ohio (1).
		Engl.: London (12); Oxford (1).
		Neth.: Amsterdam (2).
		Switz.: Geneva (1).
RUSSIAN	58	USSR: Moscow (39); Leningrad (9); Novosibirsk (1); Rostov-on-Don (1); Tomsk (1); unknown (3).
		Bel. SSR: Minsk (1).
		Latv. SSR: Riga (1).
		Lit. SSR: Vil'nus (1).
		Uzb. SSR: Tashkent (1).
GERMAN	3	Ger.: Weinheim (1).
		E. Ger.: Leipzig (1).
		USA: New York, NY (1).
POLISH	2	Pol: Poznan (1); Warsaw (1).
FRENCH	1	Fr.: Paris (1).
HUNGARIAN	1	Hung.: Budapest (1).
UKRAINIAN	1	Ukr. SSR: Kiev (1).

Author Index

Subject Index

Formula Index